Everyone's Guide to Planet Saturn

Compiled by

Nicholle Rojas

Scribbles

Year of Publication 2018

ISBN : 9789352979547

Book Published by

Scribbles

(An Imprint of Alpha Editions)

email - alphaedis@gmail.com

Produced by: PediaPress GmbH
Limburg an der Lahn
Germany
http://pediapress.com/

Contents

Introduction

Saturn

<indicator name="pp-default"> 🔒 </indicator>

Saturn

Saturn in natural color approaching equinox, photographed by *Cassini* in July 2008. The dot in the bottom left corner is Titan.

Designations	
Pronunciation	/'sætərn/ (🔊 listen)
Named after	Saturn
Adjectives	Saturnian, Cronian
Orbital characteristics	
Epoch J2000.0	
Aphelion	1,514.50 million km (10.1238 AU)
Perihelion	1,352.55 million km (9.0412 AU)
Semi-major axis	1,433.53 million km (9.5826 AU)
Eccentricity	0.0565
Orbital period	• 29.4571 yr • 10,759.22 d • 24,491.07 Saturnian solar days
Synodic period	378.09 days

Average orbital speed	9.68 km/s (6.01 mi/s)
Mean anomaly	317.020°
Inclination	• 2.485° to ecliptic • 5.51° to Sun's equator • 0.93° to invariable plane
Longitude of ascending node	113.665°
Argument of perihelion	339.392°
Known satellites	62 with formal designations; innumerable additional moonlets.
	Physical characteristics
Mean radius	58,232 km (36,184 mi)
Equatorial radius	• 60,268 km (37,449 mi)[1] • 9.449 Earths
Polar radius	• 54,364 km (33,780 mi) • 8.552 Earths
Flattening	0.09796
Surface area	• 4.27×10^{10} km^2 (1.65×10^{10} sq mi) • 83.703 Earths
Volume	• 8.2713×10^{14} km^3 (1.9844×10^{14} cu mi) • 763.59 Earths
Mass	• 5.6834×10^{26} kg (1.2530×10^{27} lb) • 95.159 Earths
Mean density	0.687 g/cm^3 (0.0248 lb/cu in)[2] (less than water)
Surface gravity	• 10.44 m/s^2 (34.3 ft/s^2) • 1.065 g
Moment of inertia factor	0.210 I/MR^2 estimate
Escape velocity	35.5 km/s (22.1 mi/s)
Sidereal rotation period	10.55 hours (10 hr 33 min)
Equatorial rotation velocity	9.87 km/s (6.13 mi/s; 35,500 km/h)
Axial tilt	26.73° (to orbit)
North pole right ascension	40.589°; 2^h 42^m 21^s
North pole declination	83.537°
Albedo	• 0.342 (Bond) • 0.499 (geometric)

Surface temp.	min	mean	max
1 bar		134 K (–139 °C)	
0.1 bar		84 K (–189 °C)	

Apparent magnitude	+1.47 to –0.24
Angular diameter	14.5" to 20.1" (excludes rings)
	Atmosphere

Surface pressure	140 kPa	
Scale height	59.5 km (37.0 mi)	
Composition by volume	by volume:	
	96.3%±2.4%	hydrogen (H$_2$)
	3.25%±2.4%	helium (He)
	0.45%±0.2%	methane (CH$_4$)
	0.0125%±0.0075%	ammonia (NH$_3$)
	0.0110%±0.0058%	hydrogen deuteride (HD)
	0.0007%±0.00015%	ethane (C$_2$H$_6$)
	Ices: • ammonia (NH$_3$) • water (H$_2$O) • ammonium hydrosulfide (NH$_4$SH)	

Saturn is the sixth planet from the Sun and the second-largest in the Solar System, after Jupiter. It is a gas giant with an average radius about nine times that of Earth. It has only one-eighth the average density of Earth, but with its larger volume Saturn is over 95 times more massive. Saturn is named after the Roman god of agriculture; its astronomical symbol (♄) represents the god's sickle.

Saturn's interior is probably composed of a core of iron–nickel and rock (silicon and oxygen compounds). This core is surrounded by a deep layer of metallic hydrogen, an intermediate layer of liquid hydrogen and liquid helium, and finally a gaseous outer layer. Saturn has a pale yellow hue due to ammonia crystals in its upper atmosphere. Electrical current within the metallic hydrogen layer is thought to give rise to Saturn's planetary magnetic field, which is weaker than Earth's, but has a magnetic moment 580 times that of Earth due to Saturn's larger size. Saturn's magnetic field strength is around one-twentieth of Jupiter's. The outer atmosphere is generally bland and lacking in contrast, although long-lived features can appear. Wind speeds on Saturn can reach 1,800 km/h (1,100 mph; 500 m/s), higher than on Jupiter, but not as high as those on Neptune.

The planet's most famous feature is its prominent ring system that is composed mostly of ice particles, with a smaller amount of rocky debris and dust. At least 62 moons are known to orbit Saturn, of which 53 are officially named. This does not include the hundreds of moonlets in the rings. Titan, Saturn's largest

Figure 1: *Composite image comparing the sizes of Saturn and Earth*

moon, and the second-largest in the Solar System, is larger than the planet Mercury, although less massive, and is the only moon in the Solar System to have a substantial atmosphere.

Physical characteristics

Saturn is a gas giant because it is predominantly composed of hydrogen and helium. It lacks a definite surface, though it may have a solid core. Saturn's rotation causes it to have the shape of an oblate spheroid; that is, it is flattened at the poles and bulges at its equator. Its equatorial and polar radii differ by almost 10%: 60,268 km versus 54,364 km. Jupiter, Uranus, and Neptune, the other giant planets in the Solar System, are also oblate but to a lesser extent. The combination of the bulge and rotation rate means that the effective surface gravity along the equator, 8.96 m/s^2, is 74% that at the poles and is lower than the surface gravity of the Earth. However, the equatorial escape velocity of nearly 36 km/s is much higher than that for the Earth.

Saturn is the only planet of the Solar System that is less dense than water—about 30% less. Although Saturn's core is considerably denser than water, the average specific density of the planet is 0.69 g/cm^3 due to the atmosphere. Jupiter has 318 times the Earth's mass, and Saturn is 95 times the mass of the

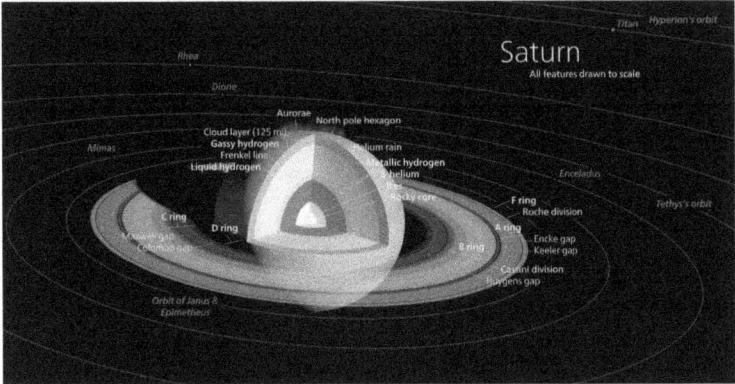

Figure 2: *Diagram of Saturn, to scale*

Earth. Together, Jupiter and Saturn hold 92% of the total planetary mass in the Solar System.

Internal structure

Despite consisting mostly of hydrogen and helium, most of Saturn's mass is not in the gas phase, because hydrogen becomes a non-ideal liquid when the density is above 0.01 g/cm^3, which is reached at a radius containing 99.9% of Saturn's mass. The temperature, pressure, and density inside Saturn all rise steadily toward the core, which causes hydrogen to be a metal in the deeper layers.

Standard planetary models suggest that the interior of Saturn is similar to that of Jupiter, having a small rocky core surrounded by hydrogen and helium with trace amounts of various volatiles. This core is similar in composition to the Earth, but more dense. Examination of Saturn's gravitational moment, in combination with physical models of the interior, has allowed constraints to be placed on the mass of Saturn's core. In 2004, scientists estimated that the core must be 9–22 times the mass of the Earth, which corresponds to a diameter of about 25,000 km. This is surrounded by a thicker liquid metallic hydrogen layer, followed by a liquid layer of helium-saturated molecular hydrogen that gradually transitions to a gas with increasing altitude. The outermost layer spans 1,000 km and consists of gas.

Saturn has a hot interior, reaching 11,700 °C at its core, and it radiates 2.5 times more energy into space than it receives from the Sun. Jupiter's thermal energy is generated by the Kelvin–Helmholtz mechanism of slow gravitational compression, but such a process alone may not be sufficient to explain heat

Figure 3: *Methane bands circle Saturn. The moon Dione hangs below the rings to the right.*

production for Saturn, because it is less massive. An alternative or additional mechanism may be generation of heat through the "raining out" of droplets of helium deep in Saturn's interior. As the droplets descend through the lower-density hydrogen, the process releases heat by friction and leaves Saturn's outer layers depleted of helium. These descending droplets may have accumulated into a helium shell surrounding the core. Rainfalls of diamonds have been suggested to occur within Saturn, as well as in Jupiter and ice giants Uranus and Neptune.

Atmosphere

The outer atmosphere of Saturn contains 96.3% molecular hydrogen and 3.25% helium by volume.[3] The proportion of helium is significantly deficient compared to the abundance of this element in the Sun. The quantity of elements heavier than helium (metallicity) is not known precisely, but the proportions are assumed to match the primordial abundances from the formation of the Solar System. The total mass of these heavier elements is estimated to be 19–31 times the mass of the Earth, with a significant fraction located in Saturn's core region.

Trace amounts of ammonia, acetylene, ethane, propane, phosphine and methane have been detected in Saturn's atmosphere. The upper clouds are

Figure 4: *A global storm girdles the planet in 2011. The head of the storm (bright area) passes the tail circling around the left limb.*

composed of ammonia crystals, while the lower level clouds appear to consist of either ammonium hydrosulfide (NH$_4$SH) or water. Ultraviolet radiation from the Sun causes methane photolysis in the upper atmosphere, leading to a series of hydrocarbon chemical reactions with the resulting products being carried downward by eddies and diffusion. This photochemical cycle is modulated by Saturn's annual seasonal cycle.

Cloud layers

Saturn's atmosphere exhibits a banded pattern similar to Jupiter's, but Saturn's bands are much fainter and are much wider near the equator. The nomenclature used to describe these bands is the same as on Jupiter. Saturn's finer cloud patterns were not observed until the flybys of the *Voyager* spacecraft during the 1980s. Since then, Earth-based telescopy has improved to the point where regular observations can be made.

The composition of the clouds varies with depth and increasing pressure. In the upper cloud layers, with the temperature in the range 100–160 K and pressures extending between 0.5–2 bar, the clouds consist of ammonia ice. Water ice clouds begin at a level where the pressure is about 2.5 bar and extend down to 9.5 bar, where temperatures range from 185–270 K. Intermixed in this layer is a band of ammonium hydrosulfide ice, lying in the pressure range 3–6 bar with temperatures of 190–235 K. Finally, the lower layers, where pressures are between 10–20 bar and temperatures are 270–330 K, contains a region of water droplets with ammonia in aqueous solution.

Saturn's usually bland atmosphere occasionally exhibits long-lived ovals and other features common on Jupiter. In 1990, the Hubble Space Telescope imaged an enormous white cloud near Saturn's equator that was not present during the *Voyager* encounters, and in 1994 another smaller storm was observed. The 1990 storm was an example of a Great White Spot, a unique but short-lived phenomenon that occurs once every Saturnian year, roughly every 30 Earth years, around the time of the northern hemisphere's summer solstice. Previous Great White Spots were observed in 1876, 1903, 1933 and 1960, with the 1933 storm being the most famous. If the periodicity is maintained, another storm will occur in about 2020.

The winds on Saturn are the second fastest among the Solar System's planets, after Neptune's. Voyager data indicate peak easterly winds of 500 m/s (1,800 km/h). In images from the *Cassini* spacecraft during 2007, Saturn's northern hemisphere displayed a bright blue hue, similar to Uranus. The color was most likely caused by Rayleigh scattering. Thermography has shown that Saturn's south pole has a warm polar vortex, the only known example of such a phenomenon in the Solar System. Whereas temperatures on Saturn are normally –185 °C, temperatures on the vortex often reach as high as –122 °C, suspected to be the warmest spot on Saturn.

North pole hexagonal cloud pattern

<templatestyles src="Multiple_image/styles.css" />

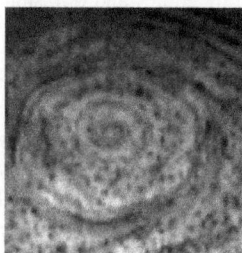

Saturn's north pole (IR animation)

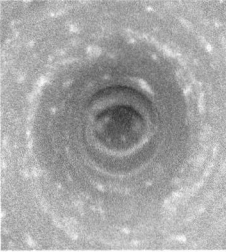

Saturn's south pole

A persisting hexagonal wave pattern around the north polar vortex in the atmosphere at about 78°N was first noted in the Voyager images. The sides of the hexagon are each about 13,800 km (8,600 mi) long, which is longer than the diameter of the Earth. The entire structure rotates with a period of 10h 39m 24s (the same period as that of the planet's radio emissions) which is assumed to be equal to the period of rotation of Saturn's interior. The hexagonal feature does not shift in longitude like the other clouds in the visible atmosphere. The pattern's origin is a matter of much speculation. Most scientists think it is a standing wave pattern in the atmosphere. Polygonal shapes have been replicated in the laboratory through differential rotation of fluids.[4,5]

South pole vortex

HST imaging of the south polar region indicates the presence of a jet stream, but no strong polar vortex nor any hexagonal standing wave. NASA reported in November 2006 that *Cassini* had observed a "hurricane-like" storm locked to the south pole that had a clearly defined eyewall. Eyewall clouds had not previously been seen on any planet other than Earth. For example, images from the *Galileo* spacecraft did not show an eyewall in the Great Red Spot of Jupiter.

The south pole storm may have been present for billions of years. This vortex is comparable to the size of Earth, and it has winds of 550 km/h.

Other features

Cassini observed a series of cloud features nicknamed "String of Pearls" found in northern latitudes. These features are cloud clearings that reside in deeper cloud layers.

Magnetosphere

<templatestyles src="Multiple_image/styles.css" />

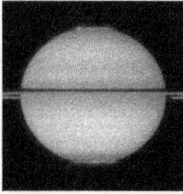

Polar aurorae on Saturn

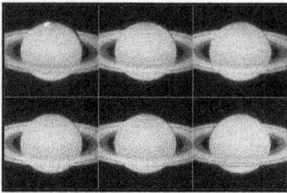

Auroral lights at Saturn's north pole

Saturn has an intrinsic magnetic field that has a simple, symmetric shape – a magnetic dipole. Its strength at the equator – 0.2 gauss (20 µT) – is approximately one twentieth of that of the field around Jupiter and slightly weaker than Earth's magnetic field. As a result, Saturn's magnetosphere is much smaller than Jupiter's. When *Voyager 2* entered the magnetosphere, the solar wind pressure was high and the magnetosphere extended only 19 Saturn radii, or 1.1 million km (712,000 mi), although it enlarged within several hours, and remained so for about three days. Most probably, the magnetic field is generated similarly to that of Jupiter – by currents in the liquid metallic-hydrogen layer called a metallic-hydrogen dynamo. This magnetosphere is efficient at deflecting the solar wind particles from the Sun. The moon Titan orbits within the outer part of Saturn's magnetosphere and contributes plasma from the ionized particles in Titan's outer atmosphere. Saturn's magnetosphere, like Earth's, produces aurorae.

Orbit and rotation

The average distance between Saturn and the Sun is over 1.4 billion kilometers (9 AU). With an average orbital speed of 9.68 km/s, it takes Saturn 10,759 Earth days (or about 29 $1/2$ years) to finish one revolution around the Sun. As a consequence, it forms a near 5:2 mean-motion resonance with Jupiter. The elliptical orbit of Saturn is inclined 2.48° relative to the orbital plane of the Earth. The perihelion and aphelion distances are, respectively, 9.195 and 9.957 AU,

Figure 5: *Saturn and rings as viewed by the Cassini spacecraft (28 October 2016)*

on average.[6] The visible features on Saturn rotate at different rates depending on latitude and multiple rotation periods have been assigned to various regions (as in Jupiter's case).

Astronomers use three different systems for specifying the rotation rate of Saturn. *System I* has a period of 10 hr 14 min 00 sec (844.3°/d) and encompasses the Equatorial Zone, the South Equatorial Belt and the North Equatorial Belt. The polar regions are considered to have rotation rates similar to *System I*. All other Saturnian latitudes, excluding the north and south polar regions, are indicated as *System II* and have been assigned a rotation period of 10 hr 38 min 25.4 sec (810.76°/d). *System III* refers to Saturn's internal rotation rate. Based on radio emissions from the planet in the period of the Voyager flybys, it has been assigned a rotation period of 10 hr 39 min 22.4 sec (810.8°/d). Because it is close to System II, it has largely superseded it.

A precise value for the rotation period of the interior remains elusive. While approaching Saturn in 2004, *Cassini* found that the radio rotation period of Saturn had increased appreciably, to approximately 10 hr 45 min 45 sec (± 36 sec). The latest estimate of Saturn's rotation (as an indicated rotation rate for Saturn as a whole) based on a compilation of various measurements from the *Cassini*, *Voyager* and Pioneer probes was reported in September 2007 is 10 hr 32 min 35 sec.

Figure 6: *A montage of Saturn and its principal moons (Dione, Tethys, Mimas, Enceladus, Rhea and Titan; Iapetus not shown). This famous image was created from photographs taken in November 1980 by the Voyager 1 spacecraft.*

In March 2007, it was found that the variation of radio emissions from the planet did not match Saturn's rotation rate. This variance may be caused by geyser activity on Saturn's moon Enceladus. The water vapor emitted into Saturn's orbit by this activity becomes charged and creates a drag upon Saturn's magnetic field, slowing its rotation slightly relative to the rotation of the planet.

An apparent oddity for Saturn is that it does not have any known trojan asteroids. These are minor planets that orbit the Sun at the stable Lagrangian points, designated L_4 and L_5, located at 60° angles to the planet along its orbit. Trojan asteroids have been discovered for Mars, Jupiter, Uranus, and Neptune. Orbital resonance mechanisms, including secular resonance, are believed to be the cause of the missing Saturnian trojans.

Natural satellites

Saturn has 62 known moons, 53 of which have formal names. In addition, there is evidence of dozens to hundreds of moonlets with diameters of 40–500 meters in Saturn's rings, which are not considered to be true moons. Titan, the largest moon, comprises more than 90% of the mass in orbit around Saturn,

Figure 7: *Possible beginning of a new moon (white dot) of Saturn (image taken by Cassini on 15 April 2013)*

including the rings. Saturn's second-largest moon, Rhea, may have a tenuous ring system of its own, along with a tenuous atmosphere.

Many of the other moons are small: 34 are less than 10 km in diameter and another 14 between 10 and 50 km in diameter. Traditionally, most of Saturn's moons have been named after Titans of Greek mythology. Titan is the only satellite in the Solar System with a major atmosphere, in which a complex organic chemistry occurs. It is the only satellite with hydrocarbon lakes.

On 6 June 2013, scientists at the IAA-CSIC reported the detection of poly-cyclic aromatic hydrocarbons in the upper atmosphere of Titan, a possible precursor for life. On 23 June 2014, NASA claimed to have strong evidence that nitrogen in the atmosphere of Titan came from materials in the Oort cloud, associated with comets, and not from the materials that formed Saturn in ear-lier times.

Saturn's moon Enceladus, which seems similar in chemical makeup to comets, has often been regarded as a potential habitat for microbial life. Evidence of this possibility includes the satellite's salt-rich particles having an "ocean-like" composition that indicates most of Enceladus's expelled ice comes from the evaporation of liquid salt water. A 2015 flyby by *Cassini* through a plume on Enceladus found most of the ingredients to sustain life forms that live by methanogenesis.

In April 2014, NASA scientists reported the possible beginning of a new moon within the A Ring, which was imaged by *Cassini* on 15 April 2013.

Planetary rings

<templatestyles src="Multiple_image/styles.css" />

The rings of Saturn (imaged here by *Cassini* in 2007) are the most massive and conspicuous in the Solar System.

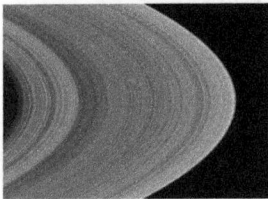

False-color UV image of Saturn's outer B and A rings; dirtier ringlets in the Cassini Division and Encke Gap show up red.

Saturn is probably best known for the system of planetary rings that makes it visually unique. The rings extend from 6,630 to 120,700 kilometers (4,120 to 75,000 mi) outward from Saturn's equator and average approximately 20 meters (66 ft) in thickness. They are composed predominantly of water ice with trace amounts of tholin impurities, and a peppered coating of approximately 7% amorphous carbon. The particles that make up the rings range in size from specks of dust up to 10 m. While the other gas giants also have ring systems, Saturn's is the largest and most visible.

There are two main hypotheses regarding the origin of the rings. One hypothesis is that the rings are remnants of a destroyed moon of Saturn. The second hypothesis is that the rings are left over from the original nebular material from which Saturn formed. Some ice in the E ring comes from the moon Enceladus's geysers. The water abundance of the rings vary radially, with the outermost ring A being the most pure in ice water. This abundance variance may be explained by meteor bombardment.

Figure 8: *Galileo Galilei first observed the rings of Saturn in 1610*

Beyond the main rings at a distance of 12 million km from the planet is the sparse Phoebe ring, which is tilted at an angle of 27° to the other rings and, like Phoebe, orbits in retrograde fashion.

Some of the moons of Saturn, including Pandora and Prometheus, act as shepherd moons to confine the rings and prevent them from spreading out. Pan and Atlas cause weak, linear density waves in Saturn's rings that have yielded more reliable calculations of their masses.

History of observation and exploration

There have been three main phases in the observation and exploration of Saturn. The first era was ancient observations (such as with the naked eye), before the invention of the modern telescopes. Starting in the 17th century progressively more advanced telescopic observations from Earth have been made. The other type is visitation by spacecraft, either by orbiting or flyby. In the 21st century observations continue from the Earth (or Earth-orbiting observatories) and from the *Cassini* orbiter at Saturn.

Ancient observations

Saturn has been known since prehistoric times and in early recorded history it was a major character in various mythologies. Babylonian astronomers systematically observed and recorded the movements of Saturn. In ancient Roman mythology, the god Saturnus, from which the planet takes its name, was the god of agriculture. The Romans considered Saturnus the equivalent of the Greek god Cronus. The Greeks had made the outermost planet sacred to Cronus, and the Romans followed suit. (In modern Greek, the planet retains its ancient name *Cronus*—Κρόνος: *Kronos*.)[7]

The Greek scientist Ptolemy based his calculations of Saturn's orbit on observations he made while it was in opposition. In Hindu astrology, there are nine astrological objects, known as Navagrahas. Saturn is known as "Shani" and judges everyone based on the good and bad deeds performed in life. Ancient Chinese and Japanese culture designated the planet Saturn as the "earth star" (土星). This was based on Five Elements which were traditionally used to classify natural elements.

In ancient Hebrew, Saturn is called 'Shabbathai'. Its angel is Cassiel. Its intelligence or beneficial spirit is 'Agîêl (Hebrew: עֲזִאֵל, translit. *'Agyal*), and its darker spirit (demon) is Zâzêl (Hebrew: זאזל translit. *Zazl*). Zazel has been described as *a great angel, invoked in Solomonic magic, who is "effective in love conjurations"*. In Ottoman Turkish, Urdu and Malay, the name of Zazel is 'Zuhal', derived from the Arabic language (Arabic: زحل, translit. *Zuhal*).

European observations (17th–19th centuries)

Saturn's rings require at least a 15-mm-diameter telescope to resolve and thus were not known to exist until Galileo first saw them in 1610. He thought of them as two moons on Saturn's sides. It was not until Christiaan Huygens used greater telescopic magnification that this notion was refuted. Huygens discovered Saturn's moon Titan; Giovanni Domenico Cassini later discovered four other moons: Iapetus, Rhea, Tethys and Dione. In 1675, Cassini discovered the gap now known as the Cassini Division.

No further discoveries of significance were made until 1789 when William Herschel discovered two further moons, Mimas and Enceladus. The irregularly shaped satellite Hyperion, which has a resonance with Titan, was discovered in 1848 by a British team.

In 1899 William Henry Pickering discovered Phoebe, a highly irregular satellite that does not rotate synchronously with Saturn as the larger moons do. Phoebe was the first such satellite found and it takes more than a year to orbit Saturn in a retrograde orbit. During the early 20th century, research on Titan led to the confirmation in 1944 that it had a thick atmosphere – a feature unique among the Solar System's moons.

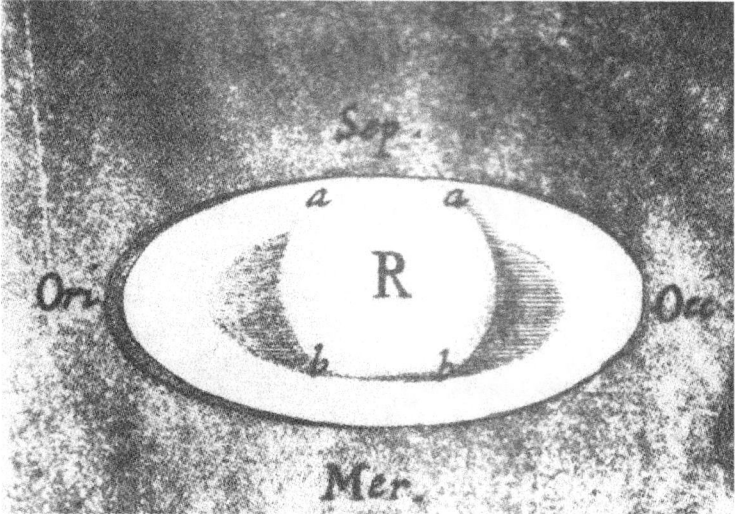

Figure 9: *Robert Hooke noted the shadows (a and b) cast by both the globe and the rings on each other in this drawing of Saturn in 1666.*

Modern NASA and ESA probes

Pioneer 11 flyby

Pioneer 11 made the first flyby of Saturn in September 1979, when it passed within 20,000 km of the planet's cloud tops. Images were taken of the planet and a few of its moons, although their resolution was too low to discern surface detail. The spacecraft also studied Saturn's rings, revealing the thin F-ring and the fact that dark gaps in the rings are bright when viewed at high phase angle (towards the Sun), meaning that they contain fine light-scattering material. In addition, Pioneer 11 measured the temperature of Titan.

Voyager flybys

In November 1980, the *Voyager 1* probe visited the Saturn system. It sent back the first high-resolution images of the planet, its rings and satellites. Surface features of various moons were seen for the first time. *Voyager 1* performed a close flyby of Titan, increasing knowledge of the atmosphere of the moon. It proved that Titan's atmosphere is impenetrable in visible wavelengths; therefore no surface details were seen. The flyby changed the spacecraft's trajectory out from the plane of the Solar System.

Almost a year later, in August 1981, *Voyager 2* continued the study of the Saturn system. More close-up images of Saturn's moons were acquired, as

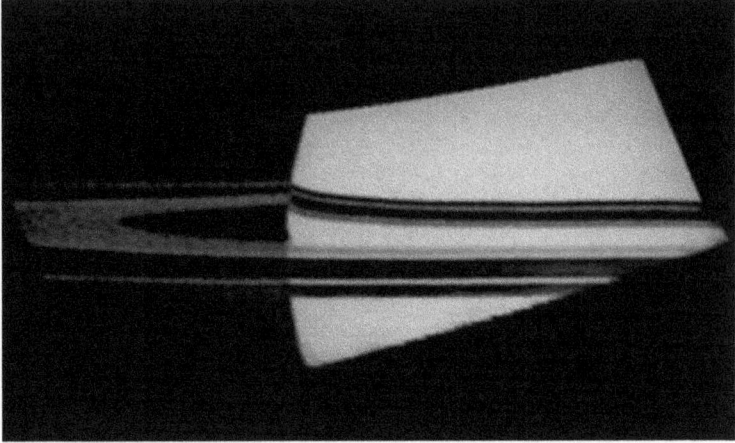

Figure 10: *Pioneer 11 image of Saturn*

well as evidence of changes in the atmosphere and the rings. Unfortunately, during the flyby, the probe's turnable camera platform stuck for a couple of days and some planned imaging was lost. Saturn's gravity was used to direct the spacecraft's trajectory towards Uranus.

The probes discovered and confirmed several new satellites orbiting near or within the planet's rings, as well as the small Maxwell Gap (a gap within the C Ring) and Keeler gap (a 42 km wide gap in the A Ring).

Cassini–Huygens spacecraft

The *Cassini–Huygens* space probe entered orbit around Saturn on 1 July 2004. In June 2004, it conducted a close flyby of Phoebe, sending back high-resolution images and data. *Cassini*'s flyby of Saturn's largest moon, Titan, captured radar images of large lakes and their coastlines with numerous islands and mountains. The orbiter completed two Titan flybys before releasing the *Huygens* probe on 25 December 2004. *Huygens* descended onto the surface of Titan on 14 January 2005.

Starting in early 2005, scientists used *Cassini* to track lightning on Saturn. The power of the lightning is approximately 1,000 times that of lightning on Earth.

In 2006, NASA reported that *Cassini* had found evidence of liquid water reservoirs no more than tens of meters below the surface that erupt in geysers on Saturn's moon Enceladus. These jets of icy particles are emitted into orbit around Saturn from vents in the moon's south polar region. Over 100 geysers have been identified on Enceladus. In May 2011, NASA scientists reported

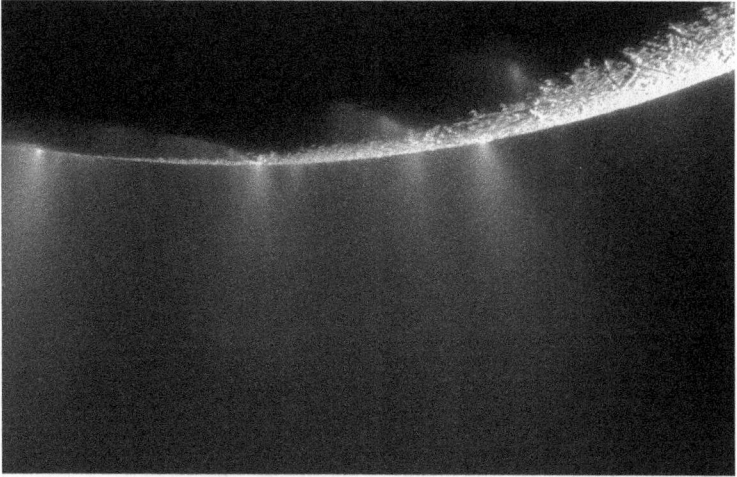

Figure 11: *At Enceladus's south pole geysers spray water from many locations along the tiger stripes.*

that Enceladus "is emerging as the most habitable spot beyond Earth in the Solar System for life as we know it".

Cassini photographs have revealed a previously undiscovered planetary ring, outside the brighter main rings of Saturn and inside the G and E rings. The source of this ring is hypothesized to be the crashing of a meteoroid off Janus and Epimetheus. In July 2006, images were returned of hydrocarbon lakes near Titan's north pole, the presence of which were confirmed in January 2007. In March 2007, hydrocarbon seas were found near the North pole, the largest of which is almost the size of the Caspian Sea. In October 2006, the probe detected an 8,000 km diameter cyclone-like storm with an eyewall at Saturn's south pole.

From 2004 to 2 November 2009, the probe discovered and confirmed eight new satellites. In April 2013 *Cassini* sent back images of a hurricane at the planet's north pole 20 times larger than those found on Earth, with winds faster than 530 km/h (330 mph). On 15 September 2017, the *Cassini-Huygens* spacecraft performed the "Grand Finale" of its mission: a number of passes through gaps between Saturn and Saturn's inner rings. The atmospheric entry of *Cassini* ended the mission.

Figure 12: *Amateur telescopic view of Saturn*

Possible future missions

The continued exploration of Saturn is still considered to be a viable option for NASA as part of their ongoing New Frontiers program of missions. NASA previously requested for plans to be put forward for a mission to Saturn that included an atmospheric entry probe and possible investigations into the habitability and possible discovery of life on Saturn's moons Titan and Enceladus.

Observation

Saturn is the most distant of the five planets easily visible to the naked eye from Earth, the other four being Mercury, Venus, Mars and Jupiter. (Uranus and occasionally 4 Vesta are visible to the naked eye in dark skies.) Saturn appears to the naked eye in the night sky as a bright, yellowish point of light with an apparent magnitude of usually between +1 and 0. It takes approximately 29.5 years for the planet to complete an entire circuit of the ecliptic against the background constellations of the zodiac. Most people will require an optical aid (very large binoculars or a small telescope) that magnifies at least 30 times to achieve an image of Saturn's rings, in which clear resolution is present. Twice every Saturnian year (roughly every 15 Earth years), the rings briefly disappear from view, due to the way in which they are angled and because they are so thin. Such a "disappearance" will next occur in 2025, but Saturn will be too close to the Sun for any ring-crossing observation to be possible.

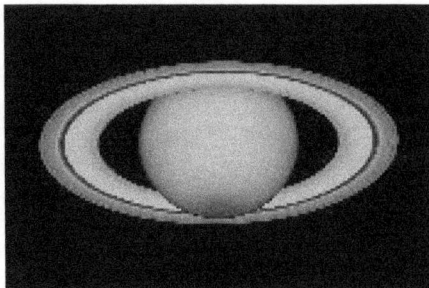

Figure 13: *Simulated appearance of Saturn as seen from Earth (at opposition) during an orbit of Saturn, 2001-2029*

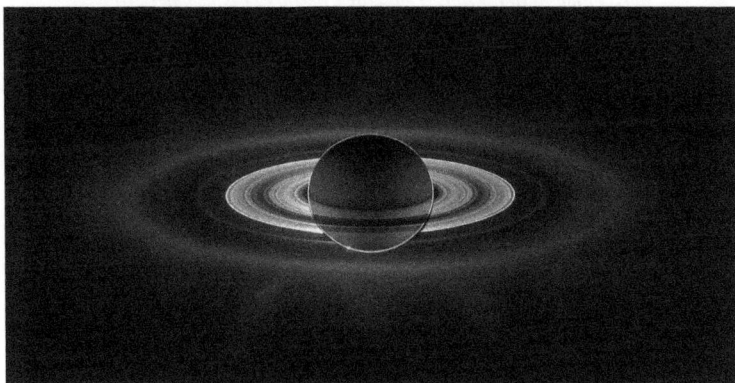

Figure 14: *Saturn eclipses the Sun, as seen from Cassini. The rings are visible, including the F Ring.*

Saturn and its rings are best seen when the planet is at, or near, opposition, the configuration of a planet when it is at an elongation of 180°, and thus appears opposite the Sun in the sky. A Saturnian opposition occurs every year—approximately every 378 days—and results in the planet appearing at its brightest. Both the Earth and Saturn orbit the Sun on eccentric orbits, which means their distances from the Sun vary over time, and therefore so do their distances from each other, hence varying the brightness of Saturn from one opposition to the next. Saturn also appears brighter when the rings are angled such that they are more visible. For example, during the opposition of 17 December 2002, Saturn appeared at its brightest due to a favorable orientation of its rings relative to the Earth, even though Saturn was closer to the Earth and Sun in late 2003.

From time to time Saturn is occulted by the Moon (that is, the Moon covers up

Figure 15:
Farewell to Saturn and moons (Enceladus, Epimetheus, Janus, Mimas, Pandora and Prometheus), by Cassini (21 November 2017).

Saturn in the sky). As with all the planets in the Solar System, occultations of Saturn occur in "seasons". Saturnian occultations will take place 12 or more times over a 12-month period, followed by about a five-year period in which no such activity is registered. Australian astronomy experts Hill and Horner explain the seasonal nature of Saturnian occultations:

> *This is the result of the fact that the moon's orbit around the Earth is tilted to the orbit of the Earth around the Sun – and so most of the time, the moon will pass above or below Saturn in the sky, and no occultation will occur. It is only when Saturn lies near the point that the moon's orbit crosses the "plane of the ecliptic" that occultations can happen – and then they occur every time the moon swings by, until Saturn moves away from the crossing point.*

Further reading

- Alexander, Arthur Francis O'Donel (1980) [1962]. *The Planet Saturn - A History of Observation, Theory and Discovery*. Dover. ISBN 978-0-486-23927-9.
- Gore, Rick (July 1981). "Voyager 1 at Saturn: Riddles of the Rings". *National Geographic*. Vol. 160 no. 1. pp. 3–31. ISSN 0027-9358[8]. OCLC 643483454[9].
- Lovett, L.; et al. (2006). *Saturn: A New View*. Harry N. Abrams. ISBN 978-0-8109-3090-2.
- Karttunen, H.; et al. (2007). *Fundamental Astronomy* (5th ed.). Springer. ISBN 978-3-540-34143-7.

- Seidelmann, P. Kenneth; et al. (2007). "Report of the IAU/IAG Working Group on cartographic coordinates and rotational elements: 2006". *Celestial Mechanics and Dynamical Astronomy*. **98** (3): 155–180. Bibcode: 2007CeMDA..98..155S[10]. doi: 10.1007/s10569-007-9072-y[11].
- de Pater, Imke; Lissauer, Jack J. (2015). *Planetary Sciences*[12] (2nd updated ed.). Cambridge University Press. p. 250. ISBN 978-0-521-85371-2.

External links

<indicator name="spoken-icon"> (◊)) </indicator>

- Saturn overview[13] by NASA's Science Mission Directorate
- Saturn fact sheet[14] at the NASA Space Science Data Coordinated Archive
- Saturnian System terminology[15] by the IAU Gazetteer of Planetary Nomenclature
- *Cassini-Huygens* legacy website[16] by the Jet Propulsion Laboratory
- Saturn[17] at SolarViews.com

<indicator name="featured-star"> ✶ </indicator>

Saturn's hexagon

Saturn's hexagon

Saturn's hexagon is a persisting hexagonal cloud pattern around the north pole of Saturn, located at about 78°N. The sides of the hexagon are about 13,800 km (8,600 mi) long, which is more than the diameter of Earth (about 12,700 km (7,900 mi)). It rotates with a period of 10h 39m 24s, the same period as Saturn's radio emissions from its interior. The hexagon does not shift in longitude like other clouds in the visible atmosphere.

Saturn's hexagon was discovered during the Voyager mission in 1981 and was later revisited by *Cassini-Huygens* in 2006. During the *Cassini* mission, the hexagon changed from a mostly blue color to more of a golden color. Saturn's south pole does not have a hexagon, according to Hubble observations; however, it does have a vortex, and there is also a vortex inside the northern hexagon. Multiple hypotheses for the hexagonal cloud pattern have been developed.

Discovery

Saturn's polar hexagon discovery was made by the Voyager mission in 1981, and it was revisited in 2006 by NASA's Cassini mission.

Cassini was able to take only thermal infrared images of the hexagon until it passed into sunlight in January 2009. Cassini was also able to take a video of the hexagonal weather pattern while traveling at the same speed as the planet, therefore recording only the movement of the hexagon. After its discovery, and after it came back into the sunlight, amateur astronomers managed to get images showing the hexagon from Earth.

Figure 16: *Saturn - North polar hexagon
and vortex as well as rings (April 2, 2014).*

Color

Between 2012 and 2016, the hexagon changed from a mostly blue color to more of a golden color. One theory for this is that sunlight is creating haze as the pole is exposed to sunlight due to the change in season. These changes were observed by the Cassini spacecraft.

Explanations for hexagon shape

One hypothesis, developed at Oxford University, is that the hexagon forms where there is a steep latitudinal gradient in the speed of the atmospheric winds in Saturn's atmosphere. Similar regular shapes were created in the laboratory when a circular tank of liquid was rotated at different speeds at its centre and periphery. The most common shape was six sided, but shapes with three to eight sides were also produced. The shapes form in an area of turbulent flow between the two different rotating fluid bodies with dissimilar speeds. A number of stable vortices of similar size form on the slower (south) side of the fluid boundary and these interact with each other to space themselves out evenly around the perimeter. The presence of the vortices influences the boundary to move northward where each is present and this gives rise to the polygon

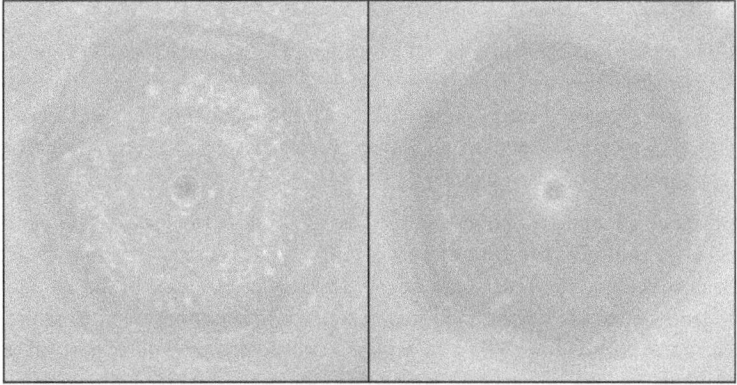

Figure 17: *2013 and 2017: hexagon color changes*

Figure 18: *False-color image from the Cassini probe*

effect. Polygons do not form at wind boundaries unless the speed differential and viscosity parameters are within certain margins and so are not present at other likely places, such as Saturn's south pole or the poles of Jupiter.

Other researchers claim that lab studies exhibit vortex streets, a series of spiraling vortices not observed in Saturn's hexagon. Simulations show that a shallow, slow, localized meandering jetstream in the same direction as Saturn's prevailing clouds is able to match the observed behaviors of Saturn's Hexagon with the same boundary stability.

Developing barotropic instability of Saturn's North Polar hexagonal circumpolar jet (Jet) plus North Polar vortex (NPV) system produces a long-living structure akin to the observed hexagon, which is not the case of the Jet-only system, which was studied in this context in a number of papers in literature. The north polar vortex (NPV), thus, plays a decisive dynamical role to stabilize hexagon jets. The influence of moist convection, which was recently suggested to be at the origin of Saturn's north polar vortex system in the literature, is investigated in the framework of the barotropic rotating shallow water model and does not alter the conclusions.

See Also

- Titanian Polar Vortices

External links

> Wikimedia Commons has media related to *Poles of Saturn*.

- Cassini Video of Saturn's Hexagon[18] on YouTube
- Saturn Revolution 175[19], Cassini images, November 27, 2012
- Saturn's Strange Hexagon – In Living Color! - Universe Today[20]
- Edge of the hexagon[21] from Planetary Photojournal[22]
- Saturn's Hexagon Comes to Light[23], APOD January 22, 2012
- In the Center of Saturn's North Polar Vortex[24], Astronomy Picture of the Day - December 4, 2012
- Video of hexagon's rotation[25] from NASA
- NASA's Cassini Spacecraft Obtains Best Views of Saturn Hexagon[26] (December 4, 2013)
- Animated vortex view[27] (TPS)
- Hexagon image[28]
- Saturn's Hexagon Replicated In Laboratory[29], video
- Hexagon Changes Color[30] (October 21, 2016)

Magnetosphere

Magnetosphere of Saturn

Magnetosphere of Saturn

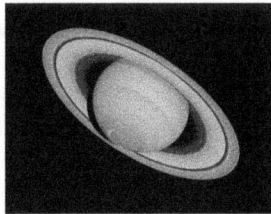

Aurorae on the south pole of Saturn as viewed by Hubble

Discovery	
Internal field	
Radius of Saturn	60,330 km
Equatorial field strength	21 μT (0.21 G)
Dipole tilt	<0.5°
Rotation period	?
Solar wind parameters	
Speed	400 km/s
IMF strength	0.5 nT
Density	0.1 cm^{-3}
Magnetospheric parameters	
Type	Intrinsic
Bow shock distance	∼27 R_s
Magnetopause distance	∼22 R_s
Main ions	O^+, H_2O^+, OH^+, H_3O^+, HO_2^+ and O_2^+ and H^+

Plasma sources	Enceladus
Mass loading rate	~100 kg/s
Maximum plasma density	50–100 cm^{-3}
Aurora	
Spectrum	radio, near-IR and UV
Total power	0.5 TW
Radio emission frequencies	10–1300 kHz

The **magnetosphere of Saturn** is the cavity created in the flow of the solar wind by the planet's internally generated magnetic field. Discovered in 1979 by the *Pioneer 11* spacecraft, Saturn's magnetosphere is the second largest of any planet in the Solar System after Jupiter. The magnetopause, the boundary between Saturn's magnetosphere and the solar wind, is located at a distance of about 20 Saturn radii from the planet's center, while its magnetotail stretches hundreds of Saturn radii behind it.

Saturn's magnetosphere is filled with plasmas originating from both the planet and its moons. The main source is the small moon Enceladus, which ejects as much as 1,000 kg/s of water vapor from the geysers on its south pole, a portion of which is ionized and forced to co-rotate with the Saturn's magnetic field. This loads the field with as much as 100 kg of water group ions per second. This plasma gradually moves out from the inner magnetosphere via the interchange instability mechanism and then escapes through the magnetotail.

The interaction between Saturn's magnetosphere and the solar wind generates bright oval aurorae around the planet's poles observed in visible, infrared and ultraviolet light. The aurorae are related to the powerful saturnian kilometric radiation (SKR), which spans the frequency interval between 100 kHz to 1300 kHz and was once thought to modulate with a period equal to the planet's rotation. However, later measurements showed that the periodicity of the SKR's modulation varies by as much as 1%, and so probably does not exactly coincide with Saturn's true rotational period, which as of 2010 remains unknown. Inside the magnetosphere there are radiation belts, which house particles with energy as high as tens of megaelectronvolts. The energetic particles have significant influence on the surfaces of inner icy moons of Saturn.

In 1980–1981 the magnetosphere of Saturn was studied by the *Voyager* spacecraft. Up until September 2017 it was a subject of ongoing investigation by Cassini mission, which arrived in 2004 and spent over 13 years observing the planet.

Discovery

Immediately after the discovery of Jupiter's decametric radio emissions in 1955, attempts were made to detect a similar emission from Saturn, but with inconclusive results.[31] The first evidence that Saturn might have an internally generated magnetic field came in 1974, with the detection of weak radio emissions from the planet at the frequency of about 1 MHz.

These medium wave emissions were modulated with a period of about 10 h 30 min, which was interpreted as Saturn's rotation period.[32] Nevertheless, the evidence available in the 1970s was too inconclusive and some scientists thought that Saturn might lack a magnetic field altogether, while others even speculated that the planet could lie beyond the heliopause.[33] The first definite detection of the saturnian magnetic field was made only on September 1, 1979, when it was passed through by the Pioneer 11 spacecraft, which measured its magnetic field strength directly.

Structure

Internal field

Like Jupiter's magnetic field, Saturn's is created by a fluid dynamo within a layer of circulating liquid metallic hydrogen in its outer core. Like Earth, Saturn's magnetic field is mostly a dipole, with north and south poles at the ends of a single magnetic axis. On Saturn, like on Jupiter, the north magnetic pole is located in the northern hemisphere, and the south magnetic pole lies in the southern hemisphere, so that magnetic field lines point away from the north pole and towards the south pole. This is reversed compared to the Earth, where the north magnetic pole lies in the southern hemisphere. Saturn's magnetic field also has quadrupole, octupole and higher components, though they are much weaker than the dipole.

The magnetic field strength at Saturn's equator is about 21 μT (0.21 G), which corresponds to a dipole magnetic moment of about 4.6×10^{18} T·m^3.[34] This makes Saturn's magnetic field slightly weaker than Earth's; however, its magnetic moment is about 580 times larger.[35] Saturn's magnetic dipole is strictly aligned with its rotational axis, meaning that the field, uniquely, is highly axisymmetric.[36] The dipole is slightly shifted (by 0.037 R_s) along Saturn's rotational axis towards the north pole.

Size and shape

Saturn's internal magnetic field deflects the solar wind, a stream of ionized particles emitted by the Sun, away from its surface, preventing it from interacting directly with its atmosphere and instead creating its own region, called a magnetosphere, composed of a plasma very different from that of the solar wind. The magnetosphere of Saturn is the second–largest magnetosphere in the Solar System after that of Jupiter.[37]

As with Earth's magnetosphere, the boundary separating the solar wind's plasma from that within Saturn's magnetosphere is called the magnetopause. The magnetopause distance from the planet's center at the subsolar point[38] varies widely from 16 to 27 R_s (R_s=60,330 km is the equatorial radius of Saturn).[39,40] The magnetopause's position depends on the pressure exerted by the solar wind, which in turn depends on solar activity. The average magnetopause standoff distance is about 22 R_s. In front of the magnetopause (at the distance of about 27 R_s from the planet)[41] lies the bow shock, a wake-like disturbance in the solar wind caused by its collision with the magnetosphere. The region between the bow shock and magnetopause is called the magnetosheath.[42]

At the opposite side of the planet, the solar wind stretches Saturn's magnetic field lines into a long, trailing magnetotail, which consists of two lobes, with the magnetic field in the northern lobe pointing away from Saturn and the southern pointing towards it. The lobes are separated by a thin layer of plasma called the tail current sheet. Like Earth's, Saturn's tail is a channel through which solar plasma enters the inner regions of the magnetosphere.[43] Similar to Jupiter, the tail is the conduit through which the plasma of the internal magnetospheric origin leaves the magnetosphere. The plasma moving from the tail to the inner magnetopshere is heated and forms a number of radiation belts.

Magnetospheric regions

Saturn's magnetosphere is often divided into four regions.[44] The innermost region co-located with Saturn's planetary rings, inside approximately 3 R_s, has a strictly dipolar magnetic field. It is largely devoid of plasma, which is absorbed by ring particles, although the radiation belts of Saturn are located in this innermost region just inside and outside the rings. The second region between 3 and 6 R_s contains the cold plasma torus and is called the inner magnetosphere. It contains the densest plasma in the saturnian system. The plasma in the torus originates from the inner icy moons and particularly from Enceladus. The magnetic field in this region is also mostly dipolar. The third region lies between 6 and 12–14 R_s and is called the dynamic and extended plasma sheet. The magnetic field in this region is stretched and non-dipolar, whereas the plasma is confined to a thin equatorial plasma sheet.[45] The fourth outermost

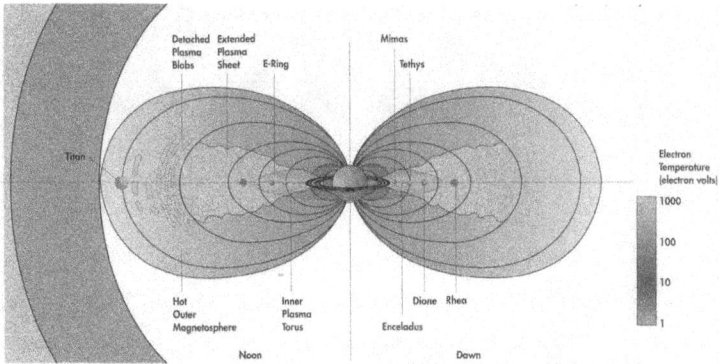

Figure 19: *The structure of Saturn's magnetosphere*

region is located beyond 15 R_s at high latitudes and continues up to magnetopause boundary. It is characterized by a low plasma density and a variable, non-dipolar magnetic field strongly influenced by the Solar wind.

In the outer parts of Saturn's magnetosphere beyond approximately 15–20 R_s[46] the magnetic field near the equatorial plane is highly stretched and forms a disk-like structure called *magnetodisk*. The disk continues up to the magnetopause on the dayside and transitions into the magnetotail on the nightside.[47] Near the dayside it can be absent when the magnetosphere is compressed by the Solar wind, which usually happens when the magnetopause distance is smaller than 23 R_s. On the nightside and flanks of the magnetosphere the magnetodisk is always present. The Saturn's magnetodisk is a much smaller analog of the Jovian magnetodisk.

The plasma sheet in the Saturn's magnetosphere has a bowl-like shape not found in any other known magnetosphere. When Cassini arrived in 2004, there was a winter in the northern hemisphere. The measurements of the magnetic field and plasma density revealed that the plasma sheet was warped and lay to the north of the equatorial plane looking like a giant bowl. Such a shape was unexpected.[48]

Dynamics

The processes driving Saturn's magnetosphere are similar to those driving Earth's and Jupiter's.[49] Just as Jupiter's magnetosphere is dominated by plasma co–rotation and mass–loading from Io, so Saturn's magnetosphere is dominated by plasma co–rotation and mass–loading from Enceladus. However, Saturn's magnetosphere is much smaller in size, while its inner region contains too

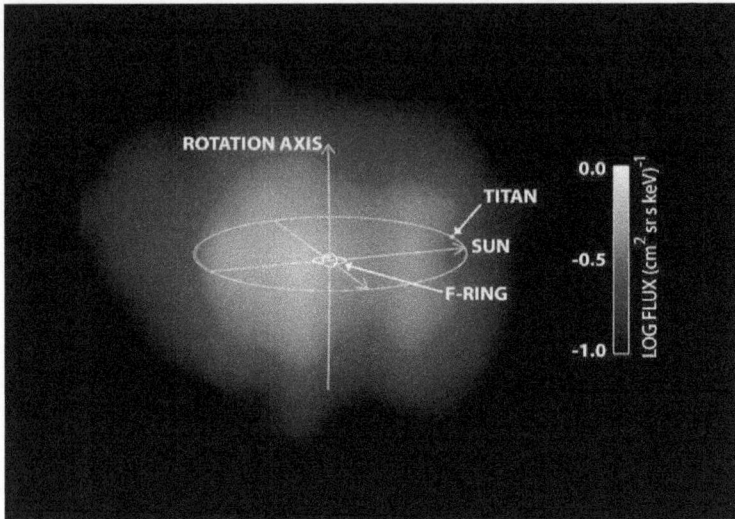

Figure 20: *Image of plasma cloud around Saturn (Cassini)*

little plasma to seriously distend it and create a large magnetodisk.[50,51] This means that it is much more strongly influenced by the solar wind, and that, like Earth's magnetic field, its dynamics are affected by reconnection with the wind similar to the Dungey cycle.

Another distinguishing feature of Saturn's magnetosphere is high abundance of neutral gas around the planet. As revealed by ultraviolet observation of Cassini, the planet is enshrouded in a large cloud of hydrogen, water vapor and their dissociative products like hydroxyl, extending as far as 45 R_s from Saturn. In the inner magnetosphere the ratio of neutrals to ions is around 60 and it increases in the outer magnetosphere, which means that the entire magnetospheric volume is filled with relatively dense weakly ionized gas. This is different, for instance, from Jupiter or Earth, where ions dominate over neutral gas, and has consequences for the magnetospheric dynamics.[52]

Sources and transport of plasma

The plasma composition in Saturn's inner magnetosphere is dominated by the water group ions: O^+, H_2O^+, OH^+ and others, hydronium ion (H_3O^+), HO_2^+ and O_2^+,[53] although protons and nitrogen ions (N^+) are also present.[54,55] The main source of water is Enceladus, which releases 300–600 kg/s of water vapor from the geysers near its south pole. The released water and hydroxyl (OH) radicals (a product of water's dissociation) form a rather thick torus around

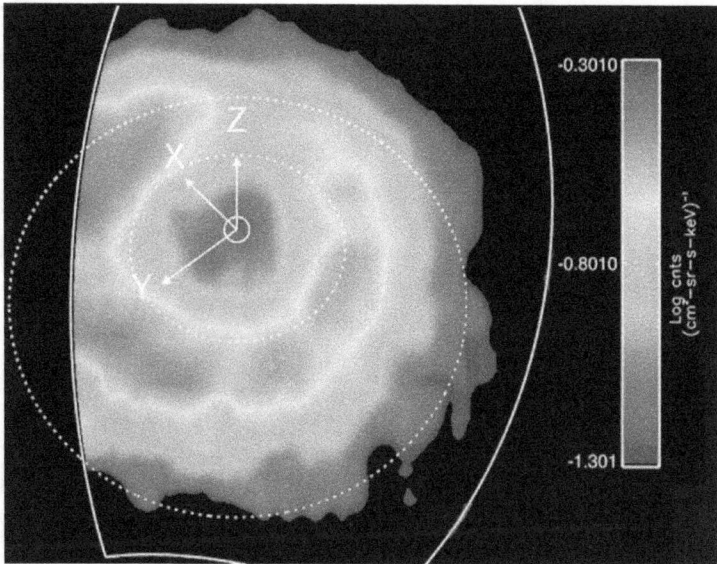

Figure 21: *Cassini image of the ring current around
Saturn carried by energetic (20–50 keV) ions*

the moon's orbit at 4 R_s with densities up to 10,000 molecules per cubic cen-
timeter.[56] At least 100 kg/s of this water is eventually ionized and added to
the co–rotating magnetospheric plasma. Additional sources of water group
ions are Saturn's rings and other icy moons.[57] The Cassini spacecraft also observed small amounts of N^+ ions in the inner magnetosphere, which probably
originate from Enceladus as well.[58]

In the outer parts of the magnetosphere the dominant ions are protons, which
originate either from the Solar wind of Saturn's ionosphere.[59] Titan, which
orbits close to the magnetopause boundary at 20 R_s, is not a significant source
of plasma.

The relatively cold plasma in the innermost region of Saturn's magnetosphere,
inside 3 R_s (near the rings) consists mainly of O^+ and O_2^+ ions. There ions
together with electrons form an ionosphere surrounding the saturnian rings.[60]

For both Jupiter and Saturn, transport of plasma from the inner to the outer
parts of the magnetosphere is thought to be related to interchange instability.
In the case of Saturn, magnetic flux tubes loaded with cold, water–rich plasma
interchange with flux tubes filled with hot plasma arriving from the outer magnetosphere. The instability is driven by centrifugal force exerted by the plasma

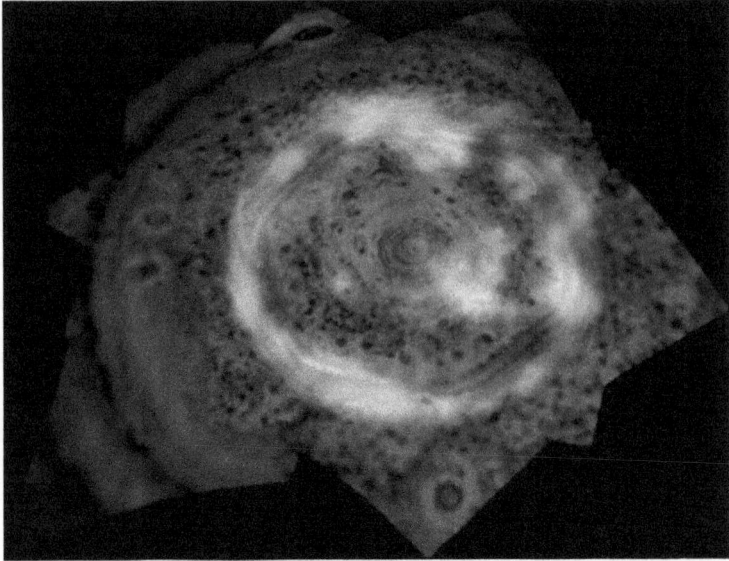

Figure 22: *The northern aurora of Saturn in the infrared light*

on the magnetic field. The cold plasma is eventually removed from the magnetosphere by plasmoids formed when the magnetic field reconnects in the magnetotail.[61] The plasmoids move down the tail and escape from the magnetosphere. The reconnection or substorm process is thought to be under the control of the solar wind and Saturn's largest moon Titan, which orbits near the outer boundary of the magnetosphere.[62]

In the magnetodisk region, beyond 6 R_s, the plasma within the co–rotating sheet exerts a significant centrifugal force on the magnetic field, causing it to stretch.[63,64] In addition, an important contribution to the ring current is provided by energetic ions with energy of more than about 10 keV.</ref> This interaction creates a current in the equatorial plane flowing azimuthally with rotation and extending as far as 20 R_s from the planet. The total strength of this current varies from 8 to 17 MA. The ring current in the saturnian magnetosphere is highly variable and depends of the solar wind pressure, being stronger when the pressure is weaker. The magnetic moment associated with this current slightly (by about 10 nT) depresses the magnetic field in the inner magnetosphere,[65] although it increases the total magnetic moment of the planet and causing the size of the magnetosphere to become larger.

Aurorae

Saturn has bright polar aurorae, which have been observed in the ultraviolet, visible and near infrared light.[66] The aurorae usually look like bright continuous circles (ovals) surrounding the poles of the planet. The latitude of auroral ovals varies in the range of 70–80°;[67] the average position is 75 ± 1° for the southern aurora, while the northern aurora is closer to the pole by about 1.5°.[68,69]</ref> From time to time either aurorae can assume a spiral shape instead of oval. In this case it begins near midnight at a latitude of around 80°, then its latitude decreases to as low as 70° as it continues into the dawn and day sectors (counterclockwise).[70] In the dusk sector the auroral latitude increases again, although when it returns to the night sector it still has a relatively low latitude and does not connect to the brighter dawn part.

Unlike Jupiter's, the Saturn's main auroral ovals are not related to the breakdown of the co–rotation of the plasma in the outer parts of the planet's magnetosphere. The aurorae on Saturn are thought to be connected to the reconnection of the magnetic field under the influence of the Solar wind (Dungey cycle), which drives an upward current (about 10 million amperes) from the ionosphere and leads to the acceleration and precipitation of energetic (1–10 keV) electrons into the polar thermosphere of Saturn.[71] The saturnian aurorae are more similar to those of the Earth, where they are also Solar wind driven. The ovals themselves correspond to the boundaries between open and closed magnetic field lines—so called polar caps, which are thought to reside at the distance of 10–15° from the poles.

The aurorae of Saturn are highly variable.[72] Their location and brightness strongly depends on the Solar wind pressure: the aurorae become brighter and move closer to the poles when the Solar wind pressure increases. The bright auroral features are observed to rotate with the angular speed of 60–75% that of Saturn. From time to time bright features appear in the dawn sector of the main oval or inside it. The average total power emitted by the aurorae is about 50 GW in the far ultraviolet (80–170 nm) and 150–300 GW in the near-infrared (3–4 μm—H_3^+ emissions) parts of the spectrum.

Saturn kilometric radiation

Saturn is the source of rather strong low frequency radio emissions called Saturn kilometric radiation (SKR). The frequency of SKR lies in the range 10–1300 kHz (wavelength of a few kilometers) with the maximum around 400 kHz. The power of these emissions is strongly modulated by the rotation of the planet and is correlated with changes in the solar wind pressure. For instance, when Saturn was immersed into the giant magnetotail of Jupiter during

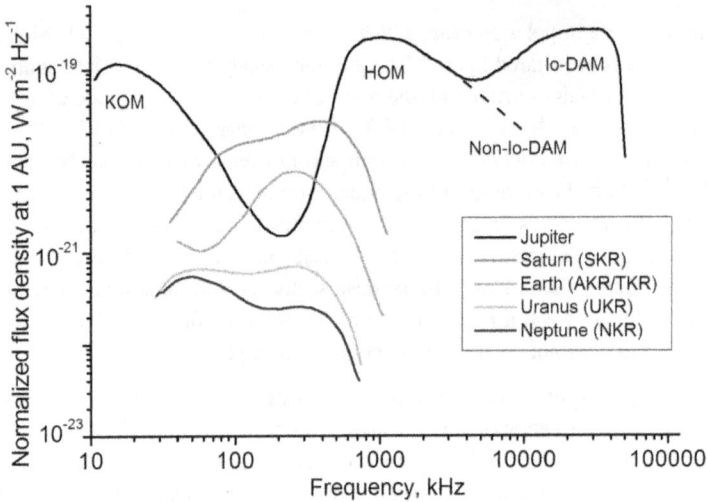

Figure 23: *The spectrum of Saturn's radio emissions compared with spectra of four other magnetized planets*

Voyager 2 flyby in 1981, the SKR power decreased greatly or even ceased completely. The kilometeric radiation is thought to be generated by the Cyclotron Maser Instability of the electrons moving along magnetic field lines related to the auroral regions of Saturn.[73] Thus the SKR is related to the auroras around the poles of the planet. The radiation itself comprises spectrally diffuse emissions as well as narrowband tones with bandwidths as narrow as 200 Hz. In the frequency–time plane arc like features are often observed, much like in the case of the Jovian kilometric radiation. The total power of the SKR is around 1 GW.[74]

The modulation of the radio emissions by planetary rotation is traditionally used to determine the rotation period of the interiors of fluid giant planets.[75] In the case of Saturn, however, this appears to be impossible, as the period varies at the timescale of tens years. In 1980–1981 the periodicity in the radio emissions as measured by Voyager 1 and 2 was 10 h 39 min 24 ± 7 s, which was then adopted as the rotational period of Saturn. Scientists were surprised when Galileo and then Cassini returned a different value—10 h 45 min 45 ± 36 s. Further observation indicated that the modulation period changes by as much as 1% on the characteristic timescale of 20–30 days with an additional long term trend. There is a correlation between the period and solar wind speed, however, the causes of this change remain a mystery. One reason may be that the saturnian perfectly axially symmetric magnetic field fails to impose a strict

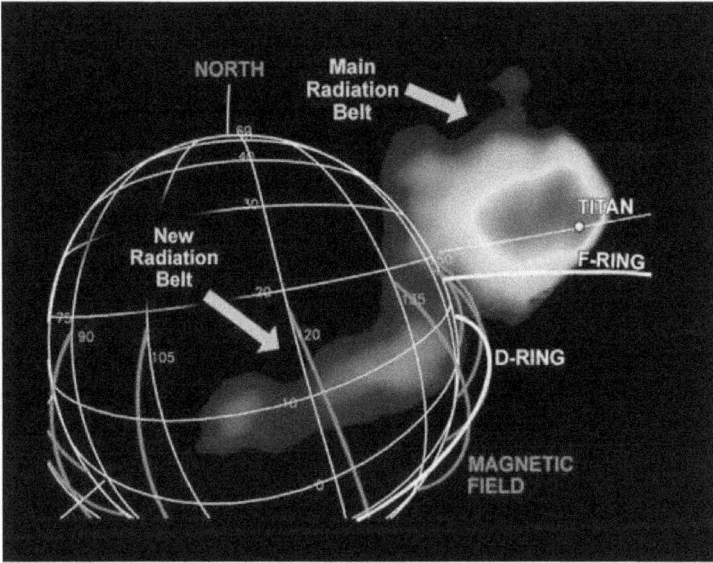

Figure 24: *Saturn's radiation belts*

corotation on the magnetospheric plasma making it slip relative to the planet. The lack of a precise correlation between the variation period of SKR and planetary rotation makes it all but impossible to determine the true rotational period of Saturn.[76]

Radiation belts

Saturn has relatively weak radiation belts, because energetic particles are absorbed by the moons and particulate material orbiting the planet. The densest (main) radiation belt lies between the inner edge of the Enceladus gas torus at 3.5 R_s and the outer edge of the A Ring at 2.3 R_s. It contains protons and relativistic electrons with energies from hundreds of kiloelectronvolts (keV) to as high as tens of megaelectronvolts (MeV) and possibly other ions. Beyond 3.5 R_s the energetic particles are absorbed by the neutral gas and their numbers drop, although less energetic particles with energies in the range of hundreds keV appear again beyond 6 R_s—these are the same particles that contribute to the ring current. The electrons in the main belt probably originate in the outer magnetosphere or Solar wind, from which they are transported by the diffusion and then adiabatically heated. However, the energetic protons consist of two populations of particles. The first population with energies of less than about 10 MeV has the same origin as electrons, while the second one with the maximum flux near 20 MeV results from the interaction of cosmic

Figure 25: *False-colour composite image showing the glow of auroras streaking out about 1,000 kilometres from the cloud tops of Saturn's south polar region*

rays with solid material present in the Saturnian system (so called cosmic ray albedo neutron decay process—CRAND). The main radiation belt of Saturn is strongly influenced by interplanetary solar wind disturbances.

The innermost region of the magnetosphere near the rings is generally devoid of energetic ions and electrons because they are absorbed by ring particles.[77] Saturn, however, has the second radiation belt discovered by Cassini in 2004 and located just inside the innermost D Ring.[78] This belt probably consists of energetic charged particles formed via the CRAND process or of ionized energetic neutral atoms coming from the main radiation belt.

The saturnian radiation belts are generally much weaker than those of Jupiter and do not emit much microwave radiation (with frequency of a few Giga-hertz). Estimates shows that their decimetric radio emissions (DIM) would be impossible to detect from the Earth.[79] Nevertherless the high energy particles cause weathering of the surfaces of the icy moons and sputter water, water products and oxygen from them.[80]

Interaction with rings and moons

The abundant population of solid bodies orbiting Saturn including moons as well as ring particles exerts a strong influence on the magnetosphere of Saturn. The plasma in the magnetosphere co-rotates with the planet, continuously impinging on the trailing hemispheres of slowly moving moons. While ring particles and the majority of moons only passively absorb plasma and energetic charged particles, three moons – Enceladus, Dione and Titan – are significant sources of new plasma.[81,82] The absorption of energetic electrons and ions reveals itself by noticeable gaps in the radiation belts of Saturn near the moon's orbits, while the dense rings of Saturn completely eliminate all energetic electrons and ions closer than 2.2 R_S, creating a low radiation zone in the vicinity of the planet. The absorption of the co-rotating plasma by a moon disturbs the magnetic field in its empty wake—the field is pulled towards a moon, creating a region of a stronger magnetic field in the near wake.[83]

The three moons mentioned above add new plasma into the magnetosphere. By far the strongest source is Enceladus, which ejects a fountain of water vapor, carbon dioxide and nitrogen through cracks in its south pole region. A fraction of this gas is ionized by the hot electrons and solar ultraviolet radiation and is added to the co-rotational plasma flow. Titan once was thought to be the principal source of plasma in Saturn's magnetosphere, especially of nitrogen. The new data obtained by Cassini in 2004–2008 established that it is not a significant source of nitrogen after all, although it may still provide significant amounts of hydrogen (due to dissociation of methane).[84] Dione is the third moon producing more new plasma than it absorbs. The mass of plasma created in the vicinity of it (about 6 g/s) is about 1/300 as much as near Enceladus. However, even this low value can not be explained only by sputtering of its icy surface by energetic particles, which may indicate that Dione is endogenically active like Enceladus. The moons that create new plasma slow the motion of the co-rotating plasma in their vicinity, which leads to the pileup of the magnetic field lines in front of them and weakening of the field in their wakes—the field drapes around them.[85] This is the opposite to what is observed for the plasma-absorbing moons.

The plasma and energetic particles present in the magnetosphere of Saturn, when absorbed by ring particles and moons, cause radiolysis of the water ice. Its products include ozone, hydrogen peroxide and molecular oxygen.[86] The first one has been detected in the surfaces of Rhea and Dione, while the second is thought to be responsible for the steep spectral slopes of moons' reflectivities in the ultraviolet region. The oxygen produced by radiolysis forms tenuous atmospheres around rings and icy moons. The ring atmosphere was detected by Cassini for the first time in 2004.[87] A fraction of the oxygen gets ionized, creating a small population of O_2^+ ions in the magnetosphere. The influence of Saturn's magnetosphere on its moons is more subtle than the influence of Jupiter on its moons. In the latter case, the magnetosphere contains a significant number of sulfur ions, which, when implanted in surfaces, produce characteristic spectral signatures. In the case of Saturn, the radiation levels are much lower and the plasma is composed mainly of water products, which, when implanted, are indistinguishable from the ice already present.

Exploration

As of 2014 the magnetosphere of Saturn has been directly explored by four spacecraft. The first mission to study the magnetosphere was Pioneer 11 in September 1979. Pioneer 11 discovered the magnetic field and made some measurements of the plasma parameters. In November 1980 and August 1981, Voyager 1–2 probes investigated the magnetosphere using an improved set of instruments. From the fly-by trajectories they measured the planetary magnetic field, plasma composition and density, high energy particle energy and spatial distribution, plasma waves and radio emissions. Cassini spacecraft was launched in 1997, and arrived in 2004, making the first measurements in more than two decades. The spacecraft continued to provide information about the magnetic field and plasma parameters of the saturnian magnetosphere until its intentional destruction on September 15, 2017.

In the 1990s, the Ulysses spacecraft conducted extensive measurements of the Saturnian kilometric radiation (SKR), which is unobservable from Earth due to the absorption in the ionosphere.[88] The SKR is powerful enough to be detected from a spacecraft at the distance of several astronomical units from the planet. Ulysses discovered that the period of the SKR varies by as much as 1%, and therefore is not directly related to the rotation period of the interior of Saturn.

Bibliography

<templatestyles src="Template:Refbegin/styles.css" />

- Andre, N.; Blanc, M.; Maurice, S.; et al. (2008). "Identification of Saturn's magnetospheric regions and associated plasma processes: Synopsis of Cassini observations during orbit insertion". *Reviews of Geophysics.* **46** (4): RG4008. Bibcode: 2008RvGeo..46.4008A[89]. doi: 10.1029/2007RG000238[90].
- Belenkaya, E.S.; Alexeev, I.I.; Kalagaev, V.V.; Blohhina, M.S. (2006). "Definition of Saturn's magnetospheric model parameters for the Pioneer 11 flyby"[91] (pdf). *Annales Geophysicae.* **24** (3): 1145–56. Bibcode: 2006AnGeo..24.1145B[92]. doi: 10.5194/angeo-24-1145-2006[93].
- Bhardwaj, Anil; Gladstone, G. Randall (2000). "Auroral emissions of the giant planets"[94] (pdf). *Reviews of Geophysics.* **38** (3): 295–353. Bibcode: 2000RvGeo..38..295B[95]. doi: 10.1029/1998RG000046[96].
- Blanc, M.; Kallenbach, R.; Erkaev, N.V. (2005). "Solar System Magnetospheres". *Space Science Reviews.* **116** (1–2): 227–298. Bibcode: 2005SSRv..116..227B[97]. doi: 10.1007/s11214-005-1958-y[98].
- Brown, Larry W. (1975). "Saturn radio emission near 1 MHz". *Journal of Geophysical Research.* **112**: L89–L92. Bibcode: 1975ApJ...198L..89B[99]. doi: 10.1086/181819[100].

- Bunce, E.J.; Cowley, S.W.H.; Alexeev, I.I.; et al. (2007). "Cassini observations of the variation of Saturn's ring current parameters with system size"[101] (pdf). *The Astrophysical Journal*. **198** (A10): A10202. Bibcode: 2007JGRA..11210202B[102]. doi: 10.1029/2007JA012275[103].

- Clark, J.T.; Gerard, J.-C.; Grodent D.; et al. (2005). "Morphological differences between Saturn's ultraviolet aurorae and those of Earth and Jupiter"[104] (PDF). *Nature*. **433** (7027): 717–719. Bibcode: 2005Natur. 433..717C[105]. doi: 10.1038/nature03331[106]. PMID 15716945[107]. Archived from the original[108] (pdf) on 2011-07-16.

- Cowley, S.W.H.; Arridge, C.S.; Bunce, E.J.; et al. (2008). "Auroral current systems in Saturn's magnetosphere: comparison of theoretical models with Cassini and HST observations"[109]. *Annales Geophysicae*. **26** (9): 2613–2630. Bibcode: 2008AnGeo..26.2613C[110]. doi: 10.5194/angeo-26-2613-2008[111].

- Gombosi, Tamas I.; Armstrong, Thomas P.; Arridge, Christopher S.; et al. (2009). "Saturn's Magnetospheric Configuration". *Saturn from Cassini–Huygens*. Springer Netherlands. pp. 203–255. doi: 10.1007/978-1-4020-9217-6_9[112]. ISBN 978-1-4020-9217-6.

- Gurnett, D.A.; Kurth, W.S.; Hospodarsky, G.B.; et al. (2005). "Radio and Plasma Wave Observations at Saturn from Cassini's Approach and First Orbit". *Science*. **307** (5713): 1255–59. Bibcode: 2005Sci... 307.1255G[113]. doi: 10.1126/science.1105356[114]. PMID 15604362[115].

- Johnson, R.E.; Luhmann, J.G.; Tokar, R.L.; et al. (2008). "Production, ionization and redistribution of O2 in Saturn's ring atmosphere"[116] (pdf). *Icarus*. **180** (2): 393–402. Bibcode: 2006Icar..180..393J[117]. doi: 10.1016/j.icarus.2005.08.021[118].

- Kivelson, Margaret Galland (2005). "The current systems of the Jovian magnetosphere and ionosphere and predictions for Saturn"[119] (pdf). *Space Science Reviews*. Springer. **116** (1–2): 299–318. Bibcode: 2005SSRv.. 116..299K[120]. doi: 10.1007/s11214-005-1959-x[121].

- Kivelson, M.G. (2005). "Transport and acceleration of plasma in the magnetospheres of Earth and Jupiter and expectations for Saturn"[122] (pdf). *Advances in Space Research*. **36** (11): 2077–89. Bibcode: 2005AdSpR..36.2077K[123]. doi: 10.1016/j.asr.2005.05.104[124].

- Kurth, W.S.; Bunce, E.J.; Clarke, J.T.; et al. (2009). "Auroral Processes". *Saturn from Cassini–Huygens*. Springer Netherlands. pp. 333–374. doi: 10.1007/978-1-4020-9217-6_12[125]. ISBN 978-1-4020-9217-6.

- Leisner, S.; Khurana, K.K.; Russell, C.T.; et al. (2007). "Observations of Enceladus and Dione as Sources for Saturn's Neutral Cloud". *Lunar and Planetary Science*. XXXVIII: 1425. Bibcode: 2007LPI....38.1425L[126].

- Mauk, B.H.; Hamilton, D.C.; Hill, T.W.; et al. (2009). "Fundamental Plasma Processes in Saturn's Magnetosphere". *Saturn from*

Cassini–Huygens. Springer Netherlands. pp. 281–331. doi: 10.1007/978-1-4020-9217-6_11[127]. ISBN 978-1-4020-9217-6.

- Nichols, J.D.; Badman, S.V.; Bunce, E.J.; et al. (2009). "Saturn's equinoctial auroras"[128] (pdf). *Geophysical Research Letters*. **36** (24): L24102:1–5. Bibcode: 2009GeoRL..3624102N[129]. doi: 10.1029/2009GL041491[130].

- Paranicas, C.; Mitchell, D.G.; Krimigis, S.M.; et al. (2007). "Sources and losses of energetic protons in Saturn's magnetosphere"[131] (pdf). *Icarus*. **197** (2): 519–525. Bibcode: 2008Icar..197..519P[132]. doi: 10.1016/j.icarus.2008.05.011[133].

- Russell, C.T. (1993). "Planetary Magnetospheres"[134] (pdf). *Reports on Progress in Physics*. **56** (6): 687–732. Bibcode: 1993RPPh...56..687R[135]. doi: 10.1088/0034-4885/56/6/001[136].

- Russell, C.T.; Jackman, C.M.; Wei, H.Y.; et al. (2008). "Titan's influence on Saturnian substorm occurrence"[137] (pdf). *Geophysical Research Letters*. **35** (12): L12105. Bibcode: 2008GeoRL..3512105R[138]. doi: 10.1029/2008GL034080[139].

- Sittler, E.C.; Andre, N.; Blanc, M.; et al. (2008). "Ion and neutral sources and sinks within Saturn's inner magnetosphere: Cassini results"[140] (pdf). *Planetary and Space Science*. **56** (1): 3–18. Bibcode: 2008P&SS...56....3S[141]. doi: 10.1016/j.pss.2007.06.006[142].

- Smith, H.T.; Shappirio, M.; Johnson, R.E.; et al. (2008). "Enceladus: A potential source of ammonia products and molecular nitrogen for Saturn's magnetosphere"[143] (pdf). *Journal of Geophysical Research*. **113** (A11): A11206. Bibcode: 2008JGRA..11311206S[144]. doi: 10.1029/2008JA013352[145].

- Smith, A.L.; Carr, T.D (1959). "Radio frequency observations of the planets in 1957–1958". *The Astrophysical Journal*. **130**: 641–647. Bibcode: 1959ApJ...130..641S[146]. doi: 10.1086/146753[147].

- Tokar, R.L.; Johnson, R.E.; Hill, T.V.; et al. (2006). "The Interaction of the Atmosphere of Enceladus with Saturn's Plasma". *Science*. **311** (5766): 1409–12. Bibcode: 2006Sci...311.1409T[148]. doi: 10.1126/science.1121061[149]. PMID 16527967[150].

- Young, D.T.; Berthelier, J.-J.; Blanc, M.; et al. (2005). "Composition and Dynamics of Plasma in Saturn's Magnetosphere". *Science*. **307** (5713): 1262–66. Bibcode: 2005Sci...307.1262Y[151]. doi: 10.1126/science.1106151[152]. PMID 15731443[153].

- Zarka, P.; Kurth, W.S. (2005). "Radio wave emissions from the outer planets before Cassini". *Space Science Reviews*. **116** (1–2): 371–397. Bibcode: 2005SSRv..116..371Z[154]. doi: 10.1007/s11214-005-1962-2[155].

- Zarka, Phillipe; Lamy, Laurent; Cecconi, Baptiste; Prangé, Renée; Rucker, Helmut O. (2007). "Modulation of Saturn's radio clock by solar wind speed"[156] (PDF). *Nature*. **450** (7167): 265–267. Bibcode: 2007Natur.450..265Z[157]. doi: 10.1038/nature06237[158]. PMID 17994092[159]. Archived from the original[160] (pdf) on 2011-06-03.

Further reading

- Arridge, C.S.; Russell, C.T.; Khurana, K.K.; et al. (2007). "Mass of Saturn's magnetodisc: Cassini observations"[161] (pdf). *Geophysical Research Letters*. **34** (9): L09108. Bibcode: 2007GeoRL..3409108A[162]. doi: 10.1029/2006GL028921[163].
- Burger, M.H.; Sittler, E.C.; Johnson, R.E.; et al. (2007). "Understanding the escape of water from Enceladus"[164] (pdf). *Journal of Geophysical Research*. **112** (A6): A06219. Bibcode: 2007JGRA..112.6219B[165]. doi: 10.1029/2006JA012086[166].
- Hill, T.W.; Thomsen, M.F.; Henderson, M.G.; et al. (2008). "Plasmoids in Saturn's magnetotail"[167] (pdf). *Journal of Geophysical Research*. **113** (A1): A01214. Bibcode: 2008JGRA..11301214H[168]. doi: 10.1029/2007JA012626[169].
- Krimigis, S.M.; Sergis, N.; Mitchell, D.G.; et al. (2007). "A dynamic, rotating ring current around Saturn"[170] (pdf). *Nature*. **450** (7172): 1050–53. Bibcode: 2007Natur.450.1050K[171]. doi: 10.1038/nature06425[172]. PMID 18075586[173].
- Martens, Hilary R.; Reisenfeld, Daniel B.; Williams, John D.; et al. (2008). "Observations of molecular oxygen ions in Saturn's inner magnetosphere"[174] (pdf). *Geophysical Research Letters*. **35** (20): L20103. Bibcode: 2008GeoRL..3520103M[175]. doi: 10.1029/2008GL035433[176].
- Russell, C.T.; Khurana, K.K.; Arridge, C.S.; Dougherty, M.K. (2008). "The magnetospheres of Jupiter and Saturn and their lessons for the Earth"[177] (pdf). *Advances in Space Research*. **41** (8): 1310–18. Bibcode: 2008AdSpR..41.1310R[178]. doi: 10.1016/j.asr.2007.07.037[179].
- Smith, H.T.; Johnson, R.E.; Sittler, E.C. (2007). "Enceladus: The likely dominant nitrogen source in Saturn's magnetosphere"[180] (pdf). *Icarus*. **188** (2): 356–366. Bibcode: 2007Icar..188..356S[181]. doi: 10.1016/j.icarus.2006.12.007[182].
- Southwood, D.J.; Kivelson, M.G. (2007). "Saturnian magnetospheric dynamics: Elucidation of a camshaft model"[183] (pdf). *Journal of Geophysical Research*. **112** (A12): A12222. Bibcode: 2007JGRA..11212222S[184]. doi: 10.1029/2007JA012254[185].
- Stallard, Tom; Miller, Steve; Melin, Henrik; et al. (2008). "Jovian-like aurorae on Saturn". *Nature*. **453** (7198): 1083–85. Bibcode: 2008Natur.453.1083S[186]. doi: 10.1038/nature07077[187]. PMID 18563160[188].
- Saturn Sends Mixed Signals[189]

External links

- NASA site about the emissions[190]

Moons of Saturn

Moons of Saturn

<templatestyles src="Multiple_image/styles.css" />

Artist's concepts of the Saturnian ring–moon system

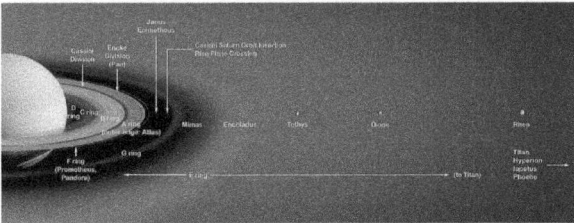

Saturn, its rings and major icy moons—from Mimas to Rhea

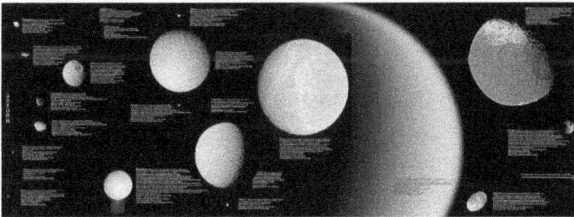

Images of several moons of Saturn. From left to right: Mimas, Enceladus, Tethys, Dione, Rhea; Titan in the background; Iapetus (top right) and irregularly shaped Hyperion (bottom right). Some small moons are also shown. All to scale.

The **moons of Saturn** are numerous and diverse, ranging from tiny moonlets less than 1 kilometer across to the enormous Titan, which is larger than the planet Mercury. Saturn has 62 moons with confirmed orbits, 53 of which have names and only 13 of which have diameters larger than 50 kilometers, as well as dense rings with complex orbital motions of their own. Seven Saturnian moons are large enough to be ellipsoidal in shape, yet only two of those, Titan and Rhea, are currently in hydrostatic equilibrium. Particularly notable among Saturn's moons are Titan, the second-largest moon (after Jupiter's Ganymede) in the Solar System, with a nitrogen-rich Earth-like atmosphere and a landscape featuring dry river networks and hydrocarbon lakes found nowhere else in the solar system; and Enceladus since its chemical composition is similar to that of comets. In particular, Enceladus emits jets of gas and dust which could indicate presence of liquid water under its south pole region and could potentially harbor a global ocean under its surface.

Twenty-four of Saturn's moons are *regular satellites*; they have prograde orbits not greatly inclined to Saturn's equatorial plane. They include the seven major satellites, four small moons that exist in a trojan orbit with larger moons, two mutually co-orbital moons and two moons that act as shepherds of Saturn's F Ring. Two other known regular satellites orbit within gaps in Saturn's rings. The relatively large Hyperion is locked in a resonance with Titan. The remaining regular moons orbit near the outer edge of the A Ring, within G Ring and between the major moons Mimas and Enceladus. The regular satellites are traditionally named after Titans and Titanesses or other figures associated with the mythological Saturn.

The remaining 38, all small except one, are *irregular satellites*, whose orbits are much farther from Saturn, have high inclinations, and are mixed between prograde and retrograde. These moons are probably captured minor planets, or debris from the breakup of such bodies after they were captured, creating collisional families. The irregular satellites have been classified by their orbital characteristics into the Inuit, Norse, and Gallic groups, and their names are chosen from the corresponding mythologies. The largest of the irregular moons is Phoebe, the ninth moon of Saturn, discovered at the end of the 19th century.

The rings of Saturn are made up of objects ranging in size from microscopic to moonlets hundreds of meters across, each in its own orbit around Saturn. Thus a precise number of Saturnian moons cannot be given, because there is no objective boundary between the countless small anonymous objects that form Saturn's ring system and the larger objects that have been named as moons. Over 150 moonlets embedded in the rings have been detected by the disturbance they create in the surrounding ring material, though this is thought to be only a small sample of the total population of such objects.

Figure 26: *Saturn (overexposed) and the moons Iapetus, Titan, Dione, Hyperion, and Rhea viewed through a 12.5-inch telescope*

Discovery

Early observations

Before the advent of telescopic photography, eight moons of Saturn were discovered by direct observation using optical telescopes. Saturn's largest moon, Titan, was discovered in 1655 by Christiaan Huygens using a 57-millimeter (2.2 in) objective lens on a refracting telescope of his own design. Tethys, Dione, Rhea and Iapetus (the "Sidera Lodoicea") were discovered between 1671 and 1684 by Giovanni Domenico Cassini. Mimas and Enceladus were discovered in 1789 by William Herschel. Hyperion was discovered in 1848 by W.C. Bond, G.P. Bond and William Lassell.

The use of long-exposure photographic plates made possible the discovery of additional moons. The first to be discovered in this manner, Phoebe, was found in 1899 by W.H. Pickering. In 1966 the tenth satellite of Saturn was discovered by Audouin Dollfus, when the rings were observed edge-on near an equinox. It was later named Janus. A few years later it was realized that all observations of 1966 could only be explained if another satellite had been present and that it had an orbit similar to that of Janus. This object is now known as Epimetheus, the eleventh moon of Saturn. It shares the same orbit with Janus—the only known example of co-orbitals in the Solar System. In 1980, three additional

Saturnian moons were discovered from the ground and later confirmed by the *Voyager* probes. They are trojan moons of Dione (Helene) and Tethys (Telesto and Calypso).

Observations by spacecraft

<templatestyles src="Multiple_image/styles.css" />

Four moons of Saturn can be seen on this image by the Cassini spacecraft: Huge Titan and Dione at the bottom, small Prometheus (under the rings) and tiny Telesto above center.

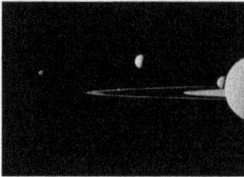

Five moons in another Cassini image: Rhea bisected in the far-right fore-ground, Mimas behind it, bright Enceladus above and beyond the rings, Pandora eclipsed by the F Ring, and Janus off to the left.

The study of the outer planets has since been revolutionized by the use of unmanned space probes. The arrival of the *Voyager* spacecraft at Saturn in 1980–1981 resulted in the discovery of three additional moons – Atlas, Prometheus and Pandora, bringing the total to 17. In addition, Epimetheus was confirmed as distinct from Janus. In 1990, Pan was discovered in archival *Voyager* images.

The *Cassini* mission, which arrived at Saturn in the summer of 2004, initially discovered three small inner moons including Methone and Pallene between Mimas and Enceladus as well as the second trojan moon of Dione – Poly-deuces. It also observed three suspected but unconfirmed moons in the F Ring. In November 2004 Cassini scientists announced that the structure of Saturn's rings indicates the presence of several more moons orbiting within the rings, although only one, Daphnis, had been visually confirmed at the time. In 2007 Anthe was announced. In 2008 it was reported that *Cassini* observations of a

Figure 27: *Quadruple Saturn–moon transit captured by the Hubble Space Telescope*

depletion of energetic electrons in Saturn's magnetosphere near Rhea might be the signature of a tenuous ring system around Saturn's second largest moon. In March 2009, Aegaeon, a moonlet within the G Ring, was announced. In July of the same year, S/2009 S 1, the first moonlet within the B Ring, was observed. In April 2014, the possible beginning of a new moon, within the A Ring, was reported. (related image)

Outer moons

Study of Saturn's moons has also been aided by advances in telescope instrumentation, primarily the introduction of digital charge-coupled devices which replaced photographic plates. For the entire 20th century, Phoebe stood alone among Saturn's known moons with its highly irregular orbit. Beginning in 2000, however, three dozen additional irregular moons have been discovered using ground-based telescopes. A survey starting in late 2000 and conducted using three medium-size telescopes found thirteen new moons orbiting Saturn at a great distance, in eccentric orbits, which are highly inclined to both the equator of Saturn and the ecliptic. They are probably fragments of larger bodies captured by Saturn's gravitational pull. In 2005, astronomers using the Mauna Kea Observatory announced the discovery of twelve more small outer moons, in 2006, astronomers using the Subaru 8.2 m telescope reported the

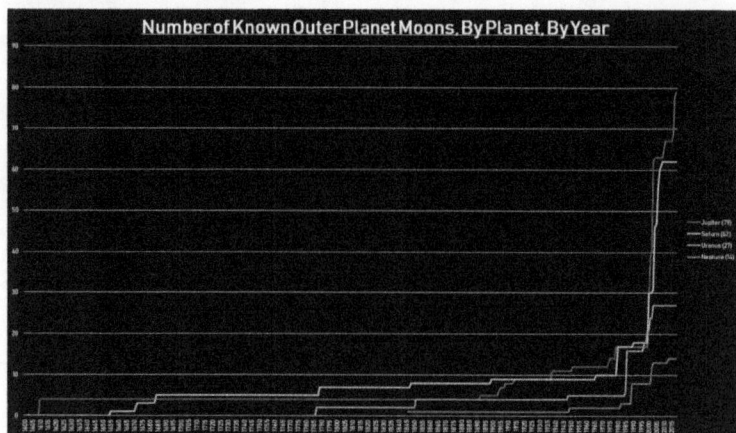

Figure 28: *The number of moons known for each of the four outer
planets up to July 2018. Saturn currently has 62 known satellites.*

discovery of nine more irregular moons, in April 2007, Tarqeq (S/2007 S 1)
was announced and in May of the same year S/2007 S 2 and S/2007 S 3 were
reported.

Some of the 62 known satellites of Saturn are considered lost because they
have not been observed since their discovery and hence their orbits are not
well-known enough to pinpoint their current locations. Work has been done
to recover many of them in surveys from 2009 onwards, but seven – S/2007
S 2, S/2004 S 13, S/2006 S 1, S/2007 S 3, S/2004 S 17, S/2004 S 12, and S/
2004 S 7 – still remain lost today.

Naming

The modern names for Saturnian moons were suggested by John Herschel in
1847. He proposed to name them after mythological figures associated with
the Roman titan of time, Saturn (equated to the Greek Cronus). In particular, the then known seven satellites were named after Titans, Titanesses and
Giants—brothers and sisters of Cronus. In 1848, Lassell proposed that the
eighth satellite of Saturn be named Hyperion after another Titan. When in
the 20th century the names of Titans were exhausted, the moons were named
after different characters of the Greco-Roman mythology or giants from other
mythologies. All the irregular moons (except Phoebe) are named after Inuit
and Gallic gods and after Norse ice giants.

Some asteroids share the same names as moons of Saturn: 55 Pandora, 106
Dione, 577 Rhea, 1809 Prometheus, 1810 Epimetheus, and 4450 Pan. In addition, two more asteroids previously shared the names of Saturnian moons until

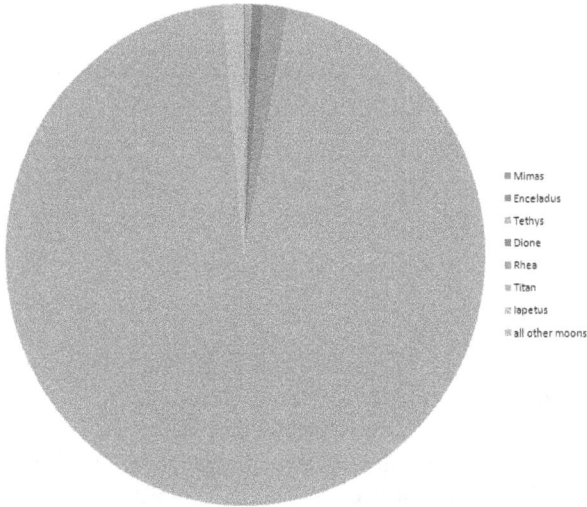

Figure 29: *The relative masses of Saturn's moons. Mimas,
the rings, and the small moons are invisible at this scale.*

spelling differences were made permanent by the International Astronomical
Union (IAU): Calypso and asteroid 53 Kalypso; and Helene and asteroid 101
Helena.

Sizes

Saturn's satellite system is very lopsided: one moon, Titan, comprises more
than 96% of the mass in orbit around the planet. The six other planemo (ellipsoidal) moons constitute roughly 4% of the mass, and the remaining 55 small
moons, together with the rings, comprise only 0.04%.[191]

Saturn's major satellites, compared to the Moon				
Name	Diameter (km)	Mass (kg)	Orbital radius (km)	Orbital period (days)
Mimas	396 (12% Moon)	4×10^{19} (0.05% Moon)	185,539 (48% Moon)	0.9 (3% Moon)
Ence-ladus	504 (14% Moon)	1.1×10^{20} (0.2% Moon)	237,948 (62% Moon)	1.4 (5% Moon)
Tethys	1,062 (30% Moon)	6.2×10^{20} (0.8% Moon)	294,619 (77% Moon)	1.9 (7% Moon)

Dione	1,123 (32% Moon)	1.1×10^{21} (1.5% Moon)	377,396 (98% Moon)	2.7 (10% Moon)
Rhea	1,527 (44% Moon)	2.3×10^{21} (3% Moon)	527,108 (137% Moon)	4.5 (20% Moon)
Titan	5,150 (148% Moon) (75% Mars)	1.35×10^{23} (180% Moon)	1,221,870 (318% Moon)	16 (60% Moon)
Iapetus	1,470 (42% Moon)	1.8×10^{21} (2.5% Moon)	3,560,820 (926% Moon)	79 (290% Moon)

Orbital groups

Although the boundaries may be somewhat vague, Saturn's moons can be divided into ten groups according to their orbital characteristics. Many of them, such as Pan and Daphnis, orbit within Saturn's ring system and have orbital periods only slightly longer than the planet's rotation period. The innermost moons and most regular satellites all have mean orbital inclinations ranging from less than a degree to about 1.5 degrees (except Iapetus, which has an inclination of 7.57 degrees) and small orbital eccentricities. On the other hand, irregular satellites in the outermost regions of Saturn's moon system, in particular the Norse group, have orbital radii of millions of kilometers and orbital periods lasting several years. The moons of the Norse group also orbit in the opposite direction to Saturn's rotation.

Ring moonlets

During late July 2009, a moonlet, S/2009 S 1, was discovered in the B Ring, 480 km from the outer edge of the ring, by the shadow it cast. It is estimated to be 300 m in diameter. Unlike the A Ring moonlets (see below), it does not induce a 'propeller' feature, probably due to the density of the B Ring.

In 2006, four tiny moonlets were found in *Cassini* images of the A Ring. Before this discovery only two larger moons had been known within gaps in the A Ring: Pan and Daphnis. These are large enough to clear continuous gaps in the ring. In contrast, a moonlet is only massive enough to clear two small—about 10 km across—partial gaps in the immediate vicinity of the moonlet itself creating a structure shaped like an airplane propeller. The moonlets themselves are tiny, ranging from about 40 to 500 meters in diameter, and are too small to be seen directly. In 2007, the discovery of 150 more moonlets revealed that they (with the exception of two that have been seen outside the Encke gap) are confined to three narrow bands in the A Ring between 126,750 and 132,000 km from Saturn's center. Each band is about a thousand kilometers

Figure 30: *Daphnis in the Keeler gap*

Figure 31: *Possible beginning of a new moon of Saturn imaged on 15 April 2014*

Figure 32: *Saturn's F Ring along with the moons, Enceladus and Rhea.*

wide, which is less than 1% the width of Saturn's rings. This region is relatively free from the disturbances caused by resonances with larger satellites, although other areas of the A Ring without disturbances are apparently free of moonlets. The moonlets were probably formed from the breakup of a larger satellite. It is estimated that the A Ring contains 7,000–8,000 propellers larger than 0.8 km in size and millions larger than 0.25 km.

Similar moonlets may reside in the F Ring. There, "jets" of material may be due to collisions, initiated by perturbations from the nearby small moon Prometheus, of these moonlets with the core of the F Ring. One of the largest F Ring moonlets may be the as-yet unconfirmed object S/2004 S 6. The F Ring also contains transient "fans" which are thought to result from even smaller moonlets, about 1 km in diameter, orbiting near the F Ring core.

One of the recently discovered moons, Aegaeon, resides within the bright arc of G Ring and is trapped in the 7:6 mean-motion resonance with Mimas. This means that it makes exactly seven revolutions around Saturn while Mimas makes exactly six. The moon is the largest among the population of bodies that are sources of dust in this ring.

In April 2014, NASA scientists reported the possible beginning of a new moon, within the A Ring.

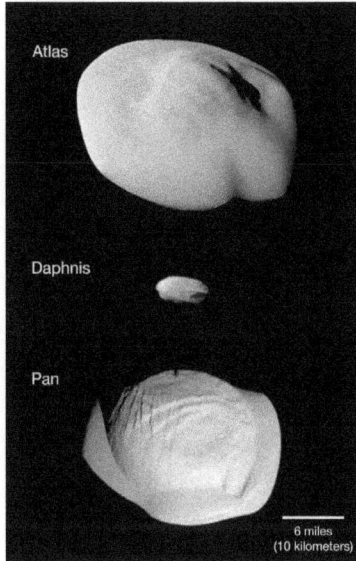

Figure 33: *Shepherd satellites – Atlas, Daphnis and Pan (color).*

Ring shepherds

Shepherd satellites are small moons that orbit within, or just beyond, a planet's ring system. They have the effect of sculpting the rings: giving them sharp edges, and creating gaps between them. Saturn's shepherd moons are Pan (Encke gap), Daphnis (Keeler gap), Atlas (A Ring), Prometheus (F Ring) and Pandora (F Ring). These moons together with co-orbitals (see below) probably formed as a result of accretion of the friable ring material on preexisting denser cores. The cores with sizes from one-third to one-half the present day moons may be themselves collisional shards formed when a parental satellite of the rings disintegrated.

Co-orbitals

Janus and Epimetheus are called co-orbital moons. They are of roughly equal size, with Janus being slightly larger than Epimetheus. Janus and Epimetheus have orbits with only a few kilometers difference in semi-major axis, close enough that they would collide if they attempted to pass each other. Instead of colliding, however, their gravitational interaction causes them to swap orbits every four years.

Figure 34: *Tiger stripes on Enceladus*

Inner large moons

<templatestyles src="Multiple_image/styles.css" />

Saturn's rings and moons

Tethys, Hyperion and Prometheus

Tethys and Janus

Tethys and the rings of Saturn

The innermost large moons of Saturn orbit within its tenuous E Ring, along with three smaller moons of the Alkyonides group.

- Mimas is the smallest and least massive of the inner round moons, although its mass is sufficient to alter the orbit of Methone. It is noticeably ovoid-shaped, having been made shorter at the poles and longer at the equator (by about 20 km) by the effects of Saturn's gravity. Mimas has a large impact crater one-third its diameter, Herschel, situated on its leading hemisphere. Mimas has no known past or present geologic activity, and its surface is dominated by impact craters. The only tectonic features known are a few arcuate and linear troughs, which probably formed when Mimas was shattered by the Herschel impact.

- Enceladus is one of the smallest of Saturn's moons that is spherical in shape—only Mimas is smaller—yet is the only small Saturnian moon that is currently endogenously active, and the smallest known body in the Solar System that is geologically active today. Its surface is morphologically diverse; it includes ancient heavily cratered terrain as well as younger smooth areas with few impact craters. Many plains on Enceladus are fractured and intersected by systems of lineaments. The area around its south pole was found by *Cassini* to be unusually warm and cut by a system of fractures about 130 km long called "tiger stripes", some of which emit jets of water vapor and dust. These jets form a large plume off its south pole, which replenishes Saturn's E ring and serves as the main source of ions in the magnetosphere of Saturn. The gas and dust are released with a rate of more than 100 kg/s. Enceladus may have liquid water underneath the south-polar surface. The source of the energy for this cryovolcanism is thought to be a 2:1 mean-motion resonance with Dione. The pure ice on the surface makes Enceladus one of the brightest known objects in the Solar System—its geometrical albedo is more than 140%.

- Tethys is the third largest of Saturn's inner moons. Its most prominent features are a large (400 km diameter) impact crater named Odysseus on its leading hemisphere and a vast canyon system named Ithaca Chasma extending at least 270° around Tethys. The Ithaca Chasma is concentric with Odysseus, and these two features may be related. Tethys appears to have no current geological activity. A heavily cratered hilly terrain occupies the majority of its surface, while a smaller and smoother plains region lies on the hemisphere opposite to that of Odysseus. The plains contain fewer craters and are apparently younger. A sharp boundary separates them from the cratered terrain. There is also a system of extensional troughs radiating away from Odysseus. The density of Tethys (0.985 g/cm^3) is less than that of water, indicating that it is made mainly of water ice with only a small fraction of rock.

- Dione is the second-largest inner moon of Saturn. It has a higher density than the geologically dead Rhea, the largest inner moon, but lower than that of active Enceladus. While the majority of Dione's surface is heavily cratered old terrain, this moon is also covered with an extensive network of troughs and lineaments, indicating that in the past it had global tectonic activity. The troughs and lineaments are especially prominent on the trailing hemisphere, where several intersecting sets of fractures form what is called "wispy terrain". The cratered plains have a few large impact craters reaching 250 km in diameter. Smooth plains with low impact-crater counts are also present on a small fraction of its surface. They were probably tectonically resurfaced relatively later in the geological history of Dione. At two locations within smooth plains strange landforms (depressions) resembling oblong impact craters have been identified, both of which lie at the centers of radiating networks of cracks and troughs; these features may be cryovolcanic in origin. Dione may be geologically active even now, although on a scale much smaller than the cryovolcanism of Enceladus. This follows from Cassini magnetic measurements that show Dione is a net source of plasma in the magnetosphere of Saturn, much like Enceladus.

Alkyonides

Three small moons orbit between Mimas and Enceladus: Methone, Anthe, and Pallene. Named after the Alkyonides of Greek mythology, they are some of the smallest moons in the Saturn system. Anthe and Methone have very faint ring arcs along their orbits, whereas Pallene has a faint complete ring. Of these three moons, only Methone has been photographed at close range, showing it to be egg-shaped with very few or no craters.[192]Template:Harvard citation documentation#Wikilink to citation does not work

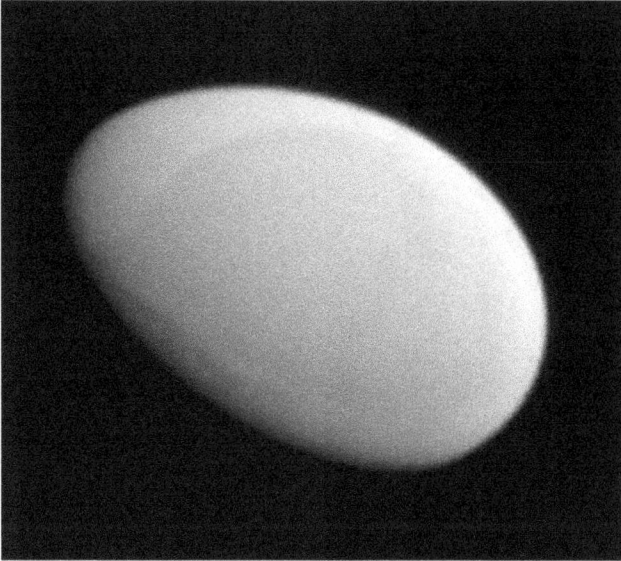

Figure 35: *Cassini image of Methone's leading side taken on 20 May 2012*

Trojan moons

Trojan moons are a unique feature only known from the Saturnian system. A trojan body orbits at either the leading L_4 or trailing L_5 Lagrange point of a much larger object, such as a large moon or planet. Tethys has two trojan moons, Telesto (leading) and Calypso (trailing), and Dione also has two, Helene (leading) and Polydeuces (trailing). Helene is by far the largest trojan moon, while Polydeuces is the smallest and has the most chaotic orbit. These moons are coated with dusty material that has smoothed out their surfaces.

Outer large moons

These moons all orbit beyond the E Ring. They are:

- Rhea is the second-largest of Saturn's moons. In 2005 *Cassini* detected a depletion of electrons in the plasma wake of Rhea, which forms when the co-rotating plasma of Saturn's magnetosphere is absorbed by the moon. The depletion was hypothesized to be caused by the presence of dust-sized particles concentrated in a few faint equatorial rings. Such a ring system would make Rhea the only moon in the Solar System known to have rings. However, subsequent targeted observations of the putative ring plane from several angles by *Cassini*'s narrow-angle camera turned up no evidence of the expected ring material, leaving the origin of the plasma

Figure 36: *Inktomi or 'The Splat', a relatively young crater with prominent butterfly-shaped ejecta on Rhea's leading hemisphere*

observations unresolved. Otherwise Rhea has rather a typical heavily cratered surface, with the exceptions of a few large Dione-type fractures (wispy terrain) on the trailing hemisphere and a very faint "line" of material at the equator that may have been deposited by material deorbiting from present or former rings. Rhea also has two very large impact basins on its anti-Saturnian hemisphere, which are about 400 and 500 km across. The first, Tirawa, is roughly comparable to the Odysseus basin on Tethys. There is also a 48 km-diameter impact crater called Inktomi[193] at 112°W that is prominent because of an extended system of bright rays, which may be one of the youngest craters on the inner moons of Saturn. No evidence of any endogenic activity has been discovered on the surface of Rhea.

- Titan, at 5,151 km diameter, is the second largest moon in the Solar System and Saturn's largest. Out of all the large moons, Titan is the only one with a dense (surface pressure of 1.5 atm), cold atmosphere, primarily made of nitrogen with a small fraction of methane. The dense atmosphere frequently produces bright white convective clouds, especially over the south pole region. On June 6, 2013, scientists at the IAA-CSIC reported the detection of polycyclic aromatic hydrocarbons in the upper atmosphere of Titan. On June 23, 2014, NASA claimed to have strong evidence that nitrogen in the atmosphere of Titan came from materials in

Figure 37: *Three crescent moons of Saturn: Titan, Mimas and Rhea*

the Oort cloud, associated with comets, and not from the materials that formed Saturn in earlier times. The surface of Titan, which is difficult to observe due to persistent atmospheric haze, shows only a few impact craters and is probably very young. It contains a pattern of light and dark regions, flow channels and possibly cryovolcanos. Some dark regions are covered by longitudinal dune fields shaped by tidal winds, where sand is made of frozen water or hydrocarbons. Titan is the only body in the Solar System beside Earth with bodies of liquid on its surface, in the form of methane–ethane lakes in Titan's north and south polar regions. The largest lake, Kraken Mare, is larger than the Caspian Sea. Like Europa and Ganymede, it is believed that Titan has a subsurface ocean made of water mixed with ammonia, which can erupt to the surface of the moon and lead to cryovolcanism. On July 2, 2014, NASA reported the ocean inside Titan may be "as salty as the Earth's Dead Sea".

- Hyperion is Titan's nearest neighbor in the Saturn system. The two moons are locked in a 4:3 mean-motion resonance with each other, meaning that while Titan makes four revolutions around Saturn, Hyperion makes exactly three. With an average diameter of about 270 km, Hyperion is smaller and lighter than Mimas. It has an extremely irregular shape, and a very odd, tan-colored icy surface resembling a sponge, though its interior may be partially porous as well. The average density of about 0.55 g/cm^3

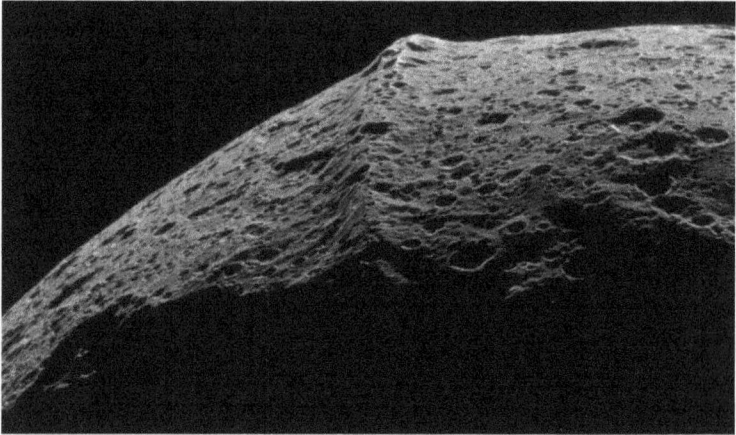

Figure 38: *Equatorial ridge on Iapetus*

indicates that the porosity exceeds 40% even assuming it has a purely icy composition. The surface of Hyperion is covered with numerous impact craters—those with diameters 2–10 km are especially abundant. It is the only moon besides the small moons of Pluto known to have a chaotic rotation, which means Hyperion has no well-defined poles or equator. While on short timescales the satellite approximately rotates around its long axis at a rate of 72–75° per day, on longer timescales its axis of rotation (spin vector) wanders chaotically across the sky. This makes the rotational behavior of Hyperion essentially unpredictable.

- Iapetus is the third-largest of Saturn's moons. Orbiting the planet at 3.5 million km, it is by far the most distant of Saturn's large moons, and also has the largest orbital inclination, at 15.47°. Iapetus has long been known for its unusual two-toned surface; its leading hemisphere is pitch-black and its trailing hemisphere is almost as bright as fresh snow. *Cassini* images showed that the dark material is confined to a large near-equatorial area on the leading hemisphere called Cassini Regio, which extends approximately from 40°N to 40°S. The pole regions of Iapetus are as bright as its trailing hemisphere. *Cassini* also discovered a 20 km tall equatorial ridge, which spans nearly the moon's entire equator. Otherwise both dark and bright surfaces of Iapetus are old and heavily cratered. The images revealed at least four large impact basins with diameters from 380 to 550 km and numerous smaller impact craters. No evidence of any endogenic activity has been discovered. A clue to the origin of the dark material covering part of Iapetus's starkly dichromatic surface may have been found in 2009, when NASA's Spitzer Space Telescope discovered a

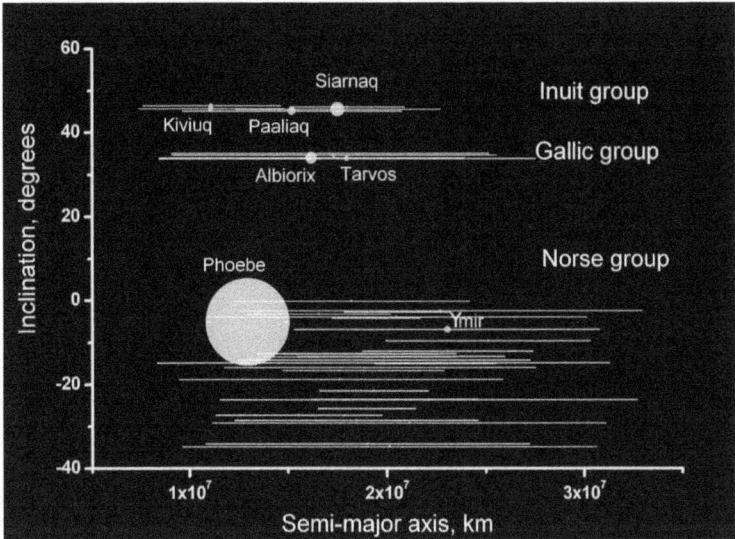

Figure 39: *Diagram illustrating the orbits of the irregular satellites of Saturn. The inclination and semi-major axis are represented on the Y and X-axis, respectively. The eccentricity of the orbits is shown by the segments extending from the pericenter to apocenter. The satellites with positive inclinations are prograde, those with negative are retrograde. The X-axis is labeled in km. The prograde Inuit and Gallic groups and the retrograde Norse group are identified.*

vast, nearly invisible disk around Saturn, just inside the orbit of the moon Phoebe – the Phoebe ring. Scientists believe that the disk originates from dust and ice particles kicked up by impacts on Phoebe. Because the disk particles, like Phoebe itself, orbit in the opposite direction to Iapetus, Iapetus collides with them as they drift in the direction of Saturn, darkening its leading hemisphere slightly. Once a difference in albedo, and hence in average temperature, was established between different regions of Iapetus, a thermal runaway process of water ice sublimation from warmer regions and deposition of water vapor onto colder regions ensued. Iapetus's present two-toned appearance results from the contrast between the bright, primarily ice-coated areas and regions of dark lag, the residue left behind after the loss of surface ice.

Irregular moons

Irregular moons are small satellites with large-radii, inclined, and frequently retrograde orbits, believed to have been acquired by the parent planet through a capture process. They often occur as collisional families or groups. The precise size as well as albedo of the irregular moons are not known for sure because the moons are very small to be resolved by a telescope, although the latter is usually assumed to be quite low—around 6% (albedo of Phoebe) or less. The irregulars generally have featureless visible and near infrared spectra dominated by water absorption bands. They are neutral or moderately red in color—similar to C-type, P-type, or D-type asteroids, though they are much less red than Kuiper belt objects.[194]

Inuit group

The Inuit group includes five prograde outer moons that are similar enough in their distances from the planet (186–297 radii of Saturn), their orbital inclinations (45–50°) and their colors that they can be considered a group. The moons are Ijiraq, Kiviuq, Paaliaq, Siarnaq, and Tarqeq. The largest among them is Siarnaq with an estimated size of about 40 km.

Gallic group

The Gallic group are four prograde outer moons that are similar enough in their distance from the planet (207–302 radii of Saturn), their orbital inclination (35–40°) and their color that they can be considered a group. They are Albiorix, Bebhionn, Erriapus, and Tarvos. Tarvos, as of 2009, is the most distant of Saturn's moons with a prograde orbit. The largest among these moons is Albiorix with an estimated size of about 32 km.

Norse group

The Norse (or Phoebe) group consists of 29 retrograde outer moons. They are Aegir, Bergelmir, Bestla, Farbauti, Fenrir, Fornjot, Greip, Hati, Hyrrokkin, Jarnsaxa, Kari, Loge, Mundilfari, Narvi, Phoebe, Skathi, Skoll, Surtur, Suttungr, Thrymr, Ymir, S/2004 S 7, S/2004 S 12, S/2004 S 13, S/2004 S 17, S/2006 S 1, S/2006 S 3, S/2007 S 2, and S/2007 S 3. After Phoebe, Ymir is the largest of the known retrograde irregular moons, with an estimated diameter of only 18 km. The Norse group may itself consist of several smaller subgroups.

- Phoebe, at 213±1.4 km in diameter, is by far the largest of Saturn's irregular satellites. It has a retrograde orbit and rotates on its axis every 9.3 hours. Phoebe was the first moon of Saturn to be studied in detail by *Cassini*, in June 2004; during this encounter *Cassini* was able to map nearly 90% of the moon's surface. Phoebe has a nearly spherical shape and a relatively high density of about 1.6 g/cm^3. *Cassini* images revealed a dark surface scarred by numerous impacts—there are about 130 craters with diameters exceeding 10 km. Spectroscopic measurement showed

Figure 40: *Saturn's rings and moons – Tethys, Enceladus and Mimas.*

that the surface is made of water ice, carbon dioxide, phyllosilicates, organics and possibly iron bearing minerals. Phoebe is believed to be a captured centaur that originated in the Kuiper belt. It also serves as a source of material for the largest known ring of Saturn, which darkens the leading hemisphere of Iapetus (see above).

List

Confirmed moons

The Saturnian moons are listed here by orbital period (or semi-major axis), from shortest to longest. Moons massive enough for their surfaces to have collapsed into a spheroid are highlighted in bold, while the irregular moons are listed in red, orange and gray background.

Key				
† Major icy moons	♠ Titan	‡ Inuit group	◇ Gallic group	♣ Norse group

Order	Label[195]	Name	Pronunciation	Image	Diameter (km)[196]	Mass (×10^15 kg)[197]	Semi-major axis (km)[198]	Orbital period (d)[199]	Inclination[200]	Eccentricity	Position	Discovery year	Discoverer
1		S/2009 S 1	—		≈ 0.3	<0.0001	≈ 117000	≈ 0.47	≈ 0°	≈ 0	outer B Ring	2009	*Cassini*
		(moonlets)	—		0.04 to 0.4 (Earhart)	<0.0001	≈ 130000	≈ 0.55	≈ 0°	≈ 0	Three 1000 km bands within A Ring	2006	*Cassini*
2	XVIII	Pan	/ˈpæn/		28.2±2.6 (34 × 31 × 20)	4.95±0.75	133584	+0.57505	0.001°	0.000035	in Encke Division	1990	M. Showalter
3	XXXV	Daphnis	/ˈdæfnɪs/		7.6±1.6 (9 × 8 × 6)	0.084±0.012	136505	+0.59408	≈ 0°	≈ 0	in Keeler Gap	2005	*Cassini*
4	XV	Atlas	/ˈætləs/		30.2±1.8 (41 × 35 × 19)	6.6±0.045	137670	+0.60169	0.003°	0.0012	outer A Ring shepherd	1980	*Voyager 2*
5	XVI	Prometheus	/prooˈmiːθiəs/-		86.2±5.4 (136 × 79 × 59)	159.5±1.5	139380	+0.61299	0.008°	0.0022	inner F Ring shepherd	1980	*Voyager 2*
6	XVII	Pandora	/pænˈdɔːrə/		81.4±3.0 (104 × 81 × 64)	137.1±1.9	141720	+0.62850	0.050°	0.0042	outer F Ring Shepherd	1980	*Voyager 2*
7a	XI	Epimetheus	/ɛpɪˈmiːθiəs/		116.2±3.6 (130 × 114 × 106)	526.6±0.6	151422	+0.69433	0.335°	0.0098	co-orbital with Janus	1977	J. Fountain, and S. Larson
7b	X	Janus	/ˈdʒeɪnəs/		179.0±2.8 (203 × 185 × 153)	1897.5±0.6	151472	+0.69466	0.165°	0.0068	co-orbital with Epimetheus	1966	A. Dollfus

9	LIII	Aegaeon	/iːˈdʒiːən/		≈ 0.5	≈ 0.0001	167500	+0.80812	0.001°	0.0002	G Ring moonlet	2008	*Cassini*
10	I	†Mimas	/ˈmaɪməs/		396.4±0.8 (416 × 393 × 381)	37493±31	185404	+0.942422	1.566°	0.0202		1789	W. Herschel
11	XXXII	Methone	/mɪˈθoʊniː/		3.2±1.2	≈ 0.02	194440	+1.00957	0.007°	0.0001	Alkyonides	2004	*Cassini*
12	XLIX	Anthe	/ˈænθiː/		1.8	≈ 0.0015	197700	+1.05089	0.1°	0.0011	Alkyonides	2007	*Cassini*
13	XXXIII	Pallene	/pəˈliːniː/		5.0±1.2 (6 × 6 × 4)	≈ 0.05	212280	+1.15375	0.181°	0.0040	Alkyonides	2004	*Cassini*
14	II	†Enceladus	/ɛnˈsɛlədəs/		504.2±0.4 (513 × 503 × 497)	108022±101	237950	+1.370218	0.010°	0.0047	Generates the E ring	1789	W. Herschel
15	III	†Tethys	/ˈtiːθɪs/		1062±1.2 (1077 × 1057 × 1053)	617449±132	294619	+1.887802	0.168°	0.0001		1684	G. Cassini
15a	XIII	Telesto	/trˈlɛstoʊ/		24.8±0.8 (33 × 24 × 20)	≈ 9.41	294619	+1.887802	1.158°	0.000	leading Tethys trojan	1980	B. Smith, H. Reitsema, S. Larson, and J. Fountain
15b	XIV	Calypso	/kəˈlɪpsoʊ/		21.4±1.4 (30 × 23 × 14)	≈ 6.3	294619	+1.887802	1.473°	0.000	trailing Tethys trojan	1980	D. Pascu, P. Seidelmann, W. Baum, and D. Currie
18	IV	†Dione	/daɪˈoʊniː/		1122.8±0.8 (1128 × 1123 × 1119)	1095452±168	377396	+2.736915	0.002°	0.0022		1684	G. Cassini
18a	XII	Helene	/ˈhɛlɪniː/		35.2±0.8 (43 × 38 × 26)	≈ 24.46	377396	+2.736915	0.212°	0.0022	leading Dione trojan	1980	P. Laques and J. Lecacheux

18b	XXXIV	Polydeuces	/ˌpɒliˈdjuːsiːz/	2.6±0.8 (3 × 2 × 1)	≈ 0.03	377396	+2.736915	0.177°	0.0192	trailing Dione trojan	2004	*Cassini* G. Cassini
21	V	†Rhea	/ˈriːə/	1527.0±1.2 (1530 × 1526 × 1525)	2306518±353	527108	+4.518212	0.327°	0.001258		1672	G. Cassini
22	VI	♠Titan	/ˈtaɪtən/	5149	134520000±20000	1221930	+15.94542	0.3485°	0.0288		1655	C. Huygens
23	VII	†Hyperion	/haɪˈpɪəriən/	270±8 (360 × 266 × 205)	5620±50	1481010	+21.27661	0.568°	0.123006	in 4:3 resonance with Titan	1848	W. Bond / G. Bond / W. Lassell
24	VIII	†Iapetus	/aɪˈæpɪtəs/	1468.6±5.6 (1491 × 1491 × 1424)	18056354±375	3560820	+79.3215	15.47°	0.028613		1671	G. Cassini
25	XXIV	‡Kiviuq	/ˈkɪviæk/	≈ 16	≈ 2.79	11294800	+448.16	49.087°	0.3288	Inuit group	2000	B. Gladman, J. Kavelaars, et al.
26	XXII	‡Ijiraq	/ˈiːrɒk/	≈ 12	≈ 1.18	11355316	+451.77	50.212°	0.3161	Inuit group	2000	B. Gladman, J. Kavelaars, et al.
27	IX	♣†Phoebe	/ˈfiːbiː/	213.0±1.4 (219 × 217 × 204)	8292±10	12869700	-545.09	173.047°	0.156242	Norse group	1899	W. Pickering
28	XX	‡Paaliaq	/ˈpɑːliɒk/	≈ 22	≈ 7.25	15103400	+692.98	46.151°	0.3631	Inuit group	2000	B. Gladman, J. Kavelaars, et al.
29	XXVII	♣Skathi	/ˈskɒði/	≈ 8	≈ 0.35	15672500	-732.52	149.084°	0.246	Norse (Skathi) Group	2000	B. Gladman, J. Kavelaars, et al.
30	XXVI	◇Albiorix	/ˌælbiˈɒrɪks/	≈ 32	≈ 22.3	16266700	+774.58	38.042°	0.477	Gallic group	2000	M. Holman

31		♣S/2007 S 2	—	≈6	≈0.15	16560000	-792.96	176.68°	0.2418	Norse group	2007	S. Sheppard, D. Jewitt, J. Kleyna, B. Marsden
32	XXXVII	◇Bebhionn	/bɛˈviːn/	≈6	≈0.15	17153520	+838.77	40.484°	0.333	Gallic group	2004	S. Sheppard, D. Jewitt, J. Kleyna
33	XXVIII	◇Erriapus	/ɛriˈæpəs/	≈10	≈0.68	17236900	+844.89	38.109°	0.4724	Gallic group	2000	B. Gladman, J. Kavelaars, et al.
34	XLVII	♣Skoll	/ˈskɒl/	≈6	≈0.15	17473800	-862.37	155.624°	0.418	Norse (Skathi) group	2006	S. Sheppard, D. Jewitt, J. Kleyna
35	XXIX	‡Siarnaq	/ˈsiːɑːrnək/	≈40	≈43.5	17776600	+884.88	45.798°	0.24961	Inuit group	2000	B. Gladman, J. Kavelaars, et al.
36	LII	‡Tarqeq	/ˈtɑːrkɛik/	≈7	≈0.23	17910600	+894.86	49.904°	0.1081	Inuit group	2007	S. Sheppard, D. Jewitt, J. Kleyna
37		♣S/2004 S 13	—	≈6	≈0.15	18056300	-905.85	167.379°	0.261	Norse group	2004	S. Sheppard, D. Jewitt, J. Kleyna
38	LI	♣Greip	/ˈgreip/	≈6	≈0.15	18065700	-906.56	172.666°	0.3735	Norse group	2006	S. Sheppard, D. Jewitt, J. Kleyna
39	XLIV	♣Hyrrokkin	/hɪˈrɒkɪn/	≈8	≈0.35	18168300	-914.29	153.272°	0.3604	Norse (Skathi) group	2006	S. Sheppard, D. Jewitt, J. Kleyna

#		Name	Pronunciation	Diameter	Albedo					Group	Year	Discoverers
40	L	♣Jarnsaxa	/jɑːrn'sæksə/	≈6	≈0.15	18556900	−943.78	162.861°	0.1918	Norse group	2006	S. Sheppard, D. Jewitt, J. Kleyna
41	XXI	◇Tarvos	/'tɑːrvəs/	≈15	≈2.3	18562800	+944.23	34.679°	0.5305	Gallic group	2000	B. Gladman, J. Kavelaars, et al.
42	XXV	♣Mundilfari /-mondel'væri/		≈7	≈0.23	18725800	−956.70	169.378°	0.198	Norse group	2000	B. Gladman, J. Kavelaars, et al.
43		♣S/2006 S 1	—	≈6	≈0.15	18930200	−972.41	154.232°	0.1303	Norse (Skathi) group	2006	S. Sheppard, D.C. Jewitt, J. Kleyna
44		♣S/2004 S 17	—	≈4	≈0.05	19099200	−985.45	166.881°	0.226	Norse group	2004	S. Sheppard, D. Jewitt, J. Kleyna
45	XXXVIII	♣Bergelmir	/beer'jelmær/	≈6	≈0.15	19104000	−985.83	157.384°	0.152	Norse (Skathi) group	2004	S. Sheppard, D. Jewitt, J. Kleyna
46	XXXI	♣Narvi	/'nɑːrvi/	≈7	≈0.23	19395200	−1008.45	137.292°	0.320	Norse (Narvi) group	2003	S. Sheppard, D. Jewitt, J. Kleyna
47	XXIII	♣Suttungr	/'sotongər/	≈7	≈0.23	19579000	−1022.82	174.321°	0.131	Norse group	2000	B. Gladman, J. Kavelaars, et al.
48	XLIII	♣Hati	/'hɑːti/	≈6	≈0.15	19709300	−1033.05	163.131°	0.291	Norse group	2004	S. Sheppard, D. Jewitt, J. Kleyna

#	Desig.	Name	Pronunciation	Diameter			Orbit	Period	Inclination	Ecc.	Group	Year	Discoverers
49		♣S/2004 S 12	—	≈5		≈0.09	19905900	−1048.54	164.042°	0.396	Norse group	2004	S. Sheppard, D. Jewitt, J. Kleyna
50	XL	♣Farbauti	/fɑːrˈbaʊti/	≈5		≈0.09	19984800	−1054.78	158.361°	0.209	Norse (Skathi) group	2004	S. Sheppard, D. Jewitt, J. Kleyna
51	XXX	♣Thrymr	/ˈθrɪmər/	≈7		≈0.23	20278100	−1078.09	174.524°	0.453	Norse group	2000	B. Gladman, J. Kavelaars, et al.
52	XXXVI	♣Aegir	/ˈaɪɪr/	≈6		≈0.15	20482900	−1094.46	167.425°	0.237	Norse group	2004	S. Sheppard, D. Jewitt, J. Kleyna
53		♣S/2007 S 3	—	≈5		≈0.09	20518500	≈ −1100	177.22°	0.130	Norse group	2007	S. Sheppard, D. Jewitt, J. Kleyna
54	XXXIX	♣Bestla	/ˈbɛstlə/	≈7		≈0.23	20570000	−1101.45	147.395°	0.5145	Norse (Narvi) group	2004	S. Sheppard, D. Jewitt, J. Kleyna
55		♣S/2004 S 7	—	≈6		≈0.15	20576700	−1101.99	165.596°	0.5299	Norse group	2004	S. Sheppard, D. Jewitt, J. Kleyna
56		♣S/2006 S 3	—	≈6		≈0.15	21076300	−1142.37	150.817°	0.4710	Norse (Skathi) group	2006	S. Sheppard, D. Jewitt, J. Kleyna
57	XLI	♣Fenrir	/ˈfɛnrɪr/	≈4		≈0.05	21930644	−1212.53	162.832°	0.131	Norse group	2004	S. Sheppard, D. Jewitt, J. Kleyna

58	XLVIII	♣Surtur	/ˈsɜːrtər/	≈ 6		≈ 0.15	22288916	–1242.36	166.918°	0.3680	Norse group	2006	S. Sheppard, D. Jewitt, J. Kleyna
59	XLV	♣Kari	/ˈkɑːri/	≈ 7		≈ 0.23	22321200	–1245.06	148.384°	0.3405	Norse (Skathi) group	2006	S. Sheppard, D. Jewitt, J. Kleyna
60	XIX	♣Ymir	/ˈiːmɪər/	≈ 18		≈ 3.97	22429673	–1254.15	172.143°	0.3349	Norse group	2000	B. Gladman, J. Kavelaars, et al.
61	XLVI	♣Loge	/ˈloʊeɪ/	≈ 6		≈ 0.15	22984322	–1300.95	166.539°	0.1390	Norse group	2006	S. Sheppard, D. Jewitt, J. Kleyna
62	XLII	♣Fornjot	/ˈfɔːrnjɒt/	≈ 6		≈ 0.15	24504879	–1432.16	167.886°	0.186	Norse group	2004	S. Sheppard, D. Jewitt, J. Kleyna

Unconfirmed moons

The following objects (observed by *Cassini*) have not been confirmed as solid bodies. It is not yet clear if these are real satellites or merely persistent clumps within the F Ring.

Name	Image	Diameter (km)	Semi-major axis (km)	Orbital period (d)	Position	Discovery year
S/2004 S 6		≈ 3–5	≈ 140130	+0.61801	uncertain objects around the F Ring	2004
S/2004 S 3/S 4[201]		≈ 3–5	≈ 140300	≈ +0.619		2004

Hypothetical moons

Two moons were claimed to be discovered by different astronomers but never seen again. Both moons were said to orbit between Titan and Hyperion.

- Chiron which was supposedly sighted by Hermann Goldschmidt in 1861, but never observed by anyone else.
- Themis was allegedly discovered in 1905 by astronomer William Pickering, but never seen again. Nevertheless, it was included in numerous almanacs and astronomy books until the 1960s.

Formation

It is thought that the Saturnian system of Titan, mid-sized moons, and rings developed from a set-up closer to the Galilean moons of Jupiter, though the details are unclear. It has been proposed either that a second Titan-sized moon broke up, producing the rings and inner mid-sized moons, or that two large moons fused to form Titan, with the collision scattering icy debris that formed the mid-sized moons.[202] On June 23, 2014, NASA claimed to have strong evidence that nitrogen in the atmosphere of Titan came from materials in the Oort cloud, associated with comets, and not from the materials that formed Saturn in earlier times. Studies based on Enceladus's tidal-based geologic activity and the lack of evidence of extensive past resonances in Tethys, Dione, and Rhea's orbits suggest that the moons inward of Titan may be only 100 million years old.

References

External links

> Wikimedia Commons has media related to *Moons of Saturn*.

- Saturn's Known Satellites[203]
- "Simulation showing the position of Saturn's Moon"[204]. Archived from the original[205] on 23 August 2011. Retrieved 26 May 2010.
- "Saturn's Rings"[206]. NASA's Solar System Exploration. Archived from the original[207] on 27 May 2010. Retrieved 26 May 2010.
- "Saturn's Moons"[208]. Astronomy Cast episode No. 61, includes full transcript. Retrieved 26 May 2010.
- Carolyn Porco. *Fly me to the moons of Saturn*[209]. Retrieved 26 May 2010.
- "The Top 10 Largest Planetary Moons"[210].
- Rotate and Spin Maps of 7 Moons[211] at *The New York Times*
- Planetary Society blog post[212] (2017-05-17) by Emily Lakdawalla with images giving comparative sizes of the moons

<indicator name="featured-star"> ⭐ </indicator>

Planetary rings

Rings of Saturn

The **rings of Saturn** are the most extensive ring system of any planet in the Solar System. They consist of countless small particles, ranging from μm to m in size, that orbit about Saturn. The ring particles are made almost entirely of water ice, with a trace component of rocky material. There is still no consensus as to their mechanism of formation; some features of the rings suggest a relatively recent origin, but theoretical models indicate they are likely to have formed early in the Solar System's history.

Although reflection from the rings increases Saturn's brightness, they are not visible from Earth with unaided vision. In 1610, the year after Galileo Galilei turned a telescope to the sky, he became the first person to observe Saturn's rings, though he could not see them well enough to discern their true nature. In 1655, Christiaan Huygens was the first person to describe them as a disk surrounding Saturn. Although many people think of Saturn's rings as being made up of a series of tiny ringlets (a concept that goes back to Laplace), true gaps are few. It is more correct to think of the rings as an annular disk with concentric local maxima and minima in density and brightness. On the scale of the clumps within the rings there is much empty space.

The rings have numerous gaps where particle density drops sharply: two opened by known moons embedded within them, and many others at locations of known destabilizing orbital resonances with the moons of Saturn. Other gaps remain unexplained. Stabilizing resonances, on the other hand, are responsible for the longevity of several rings, such as the Titan Ringlet and the G Ring.

Well beyond the main rings is the Phoebe ring, which is presumed to originate from Phoebe and thus to share its retrograde orbital motion. It is aligned with the plane of Saturn's orbit. Saturn has an axial tilt of 27 degrees, so this ring is tilted at an angle of 27 degrees to the more visible rings orbiting above Saturn's equator.

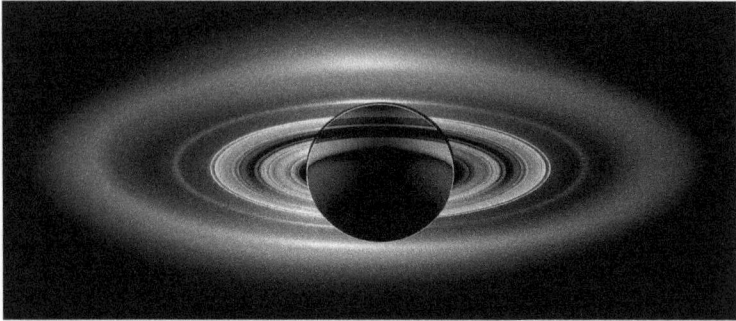

Figure 41: *The full set of rings, imaged as Saturn eclipsed the Sun from the vantage of the Cassini orbiter, 1.2 million km distant, on 19 July 2013 (brightness is exaggerated). Earth appears as a dot at 4 o'clock, between the G and E rings.*

Figure 42: *Voyager 2 view of Saturn casting a shadow across its rings. Four satellites, two of their shadows and ring spokes are visible.*

Figure 43: *Galileo first observed the rings in 1610.*

History

Galileo's work

Galileo Galilei was the first to observe the rings of Saturn in 1610 using his telescope, but was unable to identify them as such. He wrote to the Duke of Tuscany that "The planet Saturn is not alone, but is composed of three, which almost touch one another and never move nor change with respect to one another. They are arranged in a line parallel to the zodiac, and the middle one (Saturn itself) is about three times the size of the lateral ones." He also described the rings as Saturn's "ears". In 1612 the Earth passed through the plane of the rings and they became invisible. Mystified, Galileo remarked "I do not know what to say in a case so surprising, so unlooked for and so novel." He mused, "Has Saturn swallowed his children?" — referring to the myth of the Titan Saturn devouring his offspring to forestall the prophecy of them over-throwing him. He was further confused when the rings again became visible in 1613.

Early astronomers used anagrams as a form of commitment scheme to lay claim to new discoveries before their results were ready for publication. Galileo used *smaismrmilmepoetaleumibunenugttauiras* for *Altissimum plan-etam tergeminum observavi* ("I have observed the most distant planet to have a triple form") for discovering the rings of Saturn.

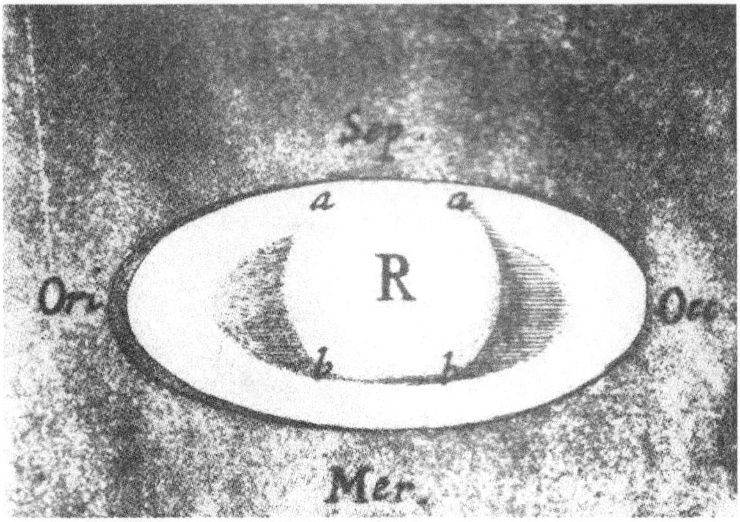

Figure 44: *Robert Hooke noted the shadows (a and b) cast by both the globe and the rings on each other in this 1666 drawing of Saturn.*

Ring theory, observations and exploration

In 1655, Christiaan Huygens became the first person to suggest that Saturn was surrounded by a ring. Using a $50\times$ power refracting telescope that he designed himself, far superior to those available to Galileo, Huygens observed Saturn and in 1656, like Galileo, had published an anagram saying "aaaaaaac-ccccdeeeeeghiiiiiiiilllllmmnnnnnnnnnnnnooooppqrrsttttttuuuuu". Upon confirming his observations, three years later he revealed it to mean "Annuto cingitur, tenui, plano, nusquam coherente, ad eclipticam inclinato"; that is, "It [Saturn] is surrounded by a thin, flat, ring, nowhere touching, inclined to the ecliptic". Robert Hooke was another early observer of the rings of Saturn, and noted the casting of shadows on the rings.

In 1675, Giovanni Domenico Cassini determined that Saturn's ring was composed of multiple smaller rings with gaps between them; the largest of these gaps was later named the Cassini Division. This division is a 4,800-km-wide region between the A ring and B Ring.

In 1787, Pierre-Simon Laplace proved that a uniform solid ring would be unstable and suggested that the rings were composed of a large number of solid ringlets.

In 1859, James Clerk Maxwell demonstrated that a nonuniform solid ring, solid ringlets or a continuous fluid ring would also not be stable, indicating that the ring must be composed of numerous small particles, all independently orbiting Saturn. Later, Sofia Kovalevskaya also found that Saturn's rings cannot be liquid ring-shaped bodies. Spectroscopic studies of the rings carried out in 1895 by James Keeler of Allegheny Observatory and Aristarkh Belopolsky of Pulkovo Observatory showed Maxwell's analysis was correct.

Four robotic spacecraft have observed Saturn's rings from the vicinity of the planet. Pioneer 11's closest approach to Saturn occurred in September 1979 at a distance of 20,900 km. Pioneer 11 was responsible for the discovery of the F ring. Voyager 1's closest approach occurred in November 1980 at a distance of 64,200 km. A failed photopolarimeter prevented Voyager 1 from observing Saturn's rings at the planned resolution; nevertheless, images from the spacecraft provided unprecedented detail of the ring system and revealed the existence of the G ring. Voyager 2's closest approach occurred in August 1981 at a distance of 41,000 km. Voyager 2's working photopolarimeter allowed it to observe the ring system at higher resolution than Voyager 1, and to thereby discover many previously unseen ringlets. *Cassini* spacecraft entered into orbit around Saturn in July 2004. *Cassini*'s images of the rings are the most detailed to-date, and are responsible for the discovery of yet more ringlets.

The rings are named alphabetically in the order they were discovered. The main rings are, working outward from the planet, C, B and A, with the Cassini Division, the largest gap, separating Rings B and A. Several fainter rings were discovered more recently. The D Ring is exceedingly faint and closest to the planet. The narrow F Ring is just outside the A Ring. Beyond that are two far fainter rings named G and E. The rings show a tremendous amount of structure on all scales, some related to perturbations by Saturn's moons, but much unexplained.

Physical characteristics

The dense main rings extend from 7,000 km (4,300 mi) to 80,000 km (50,000 mi) away from Saturn's equator, whose radius is 60,300 km (37,500 mi) (see Major subdivisions). With an estimated local thickness of as little as 10 m and as much as 1 km, they are composed of 99.9% pure water ice with a smattering of impurities that may include tholins or silicates. The main rings are primarily composed of particles ranging in size from 1 cm to 10 m.

Based on *Voyager* observations, the total mass of the rings was estimated to be about 3×10^{19} kg. This is a small fraction of the total mass of Saturn (about 50 ppb) and is just a little less than the moon Mimas.[213] More recent observations and computer modeling based on *Cassini* observations show that

Figure 45: *Simulated image using color to present radio-occultation-derived particle size data. The attenuation of 0.94-, 3.6-, and 13-cm signals sent by Cassini through the rings to Earth shows abundance of particles of sizes similar to or larger than those wavelengths. Purple (B, inner A Ring) means few particles are < 5 cm (all signals similarly attenuated). Green and blue (C, outer A Ring) mean particles < 5 cm and < 1 cm, respectively, are common. White areas (B Ring) are too dense to transmit adequate signal. Other evidence shows rings A to C have a broad range of particle sizes, up to m across.*

this may be an underestimate due to clumping in the rings and the mass may be three times this figure. Although the largest gaps in the rings, such as the Cassini Division and Encke Gap, can be seen from Earth, both *Voyager* spacecraft discovered that the rings have an intricate structure of thousands of thin gaps and ringlets. This structure is thought to arise, in several different ways, from the gravitational pull of Saturn's many moons. Some gaps are cleared out by the passage of tiny moonlets such as Pan, many more of which may yet be discovered, and some ringlets seem to be maintained by the gravitational effects of small shepherd satellites (similar to Prometheus and Pandora's maintenance of the F ring).Wikipedia:Citation needed Other gaps arise from resonances between the orbital period of particles in the gap and that of a more massive moon further out; Mimas maintains the Cassini Division in this manner. Still more structure in the rings consists of spiral waves raised by the inner moons' periodic gravitational perturbations at less disruptive reso-

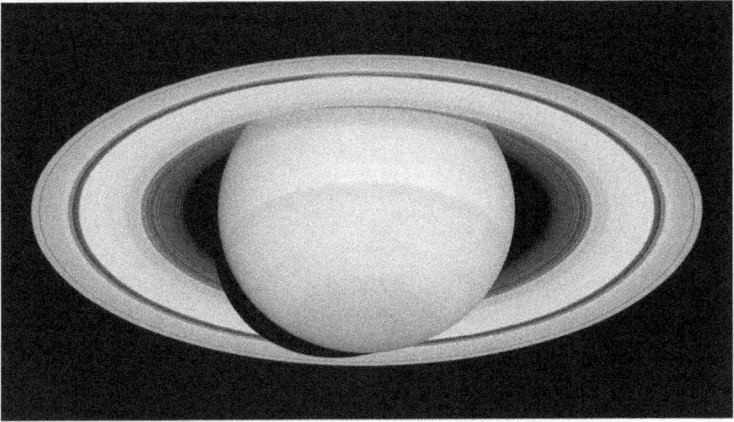

Figure 46: *The dark Cassini Division separates the wide inner B Ring and outer A ring in this image from the HST's ACS (March 22, 2004). The less prominent C Ring is just inside the B Ring.*

Figure 47: *Cassini mosaic of Saturn's rings on August 12, 2009, a day after equinox. With the rings pointed at the Sun, illumination is by light reflected off Saturn, except on thicker or out-of-plane sections, like the F Ring.*

Figure 48: *Cassini space probe view of the unilluminated side of Saturn's rings (May 9, 2007).*

nances.Wikipedia:Citation needed Data from the *Cassini* space probe indicate
that the rings of Saturn possess their own atmosphere, independent of that of
the planet itself. The atmosphere is composed of molecular oxygen gas (O_2)
produced when ultraviolet light from the Sun interacts with water ice in the
rings. Chemical reactions between water molecule fragments and further ultraviolet stimulation create and eject, among other things, O_2. According to
models of this atmosphere, H_2 is also present. The O_2 and H_2 atmospheres
are so sparse that if the entire atmosphere were somehow condensed onto the
rings, it would be about one atom thick. The rings also have a similarly sparse
OH (hydroxide) atmosphere. Like the O_2, this atmosphere is produced by
the disintegration of water molecules, though in this case the disintegration
is done by energetic ions that bombard water molecules ejected by Saturn's
moon Enceladus. This atmosphere, despite being extremely sparse, was detected from Earth by the Hubble Space Telescope. Saturn shows complex
patterns in its brightness. Most of the variability is due to the changing aspect
of the rings, and this goes through two cycles every orbit. However, super-
imposed on this is variability due to the eccentricity of the planet's orbit that
causes the planet to display brighter oppositions in the northern hemisphere
than it does in the southern.

In 1980, *Voyager 1* made a fly-by of Saturn that showed the F-ring to be composed of three narrow rings that appeared to be braided in a complex structure;
it is now known that the outer two rings consist of knobs, kinks and lumps

that give the illusion of braiding, with the less bright third ring lying inside them.Wikipedia:Citation needed

New images of the rings taken around the 11 August 2009 equinox of Saturn by NASA's *Cassini* spacecraft have shown that the rings extend significantly out of the nominal ring plane in a few places. This displacement reaches as much as 4 km (2.5 mi) at the border of the Keeler Gap, due to the out-of-plane orbit of Daphnis, the moon that creates the gap.

Formation of main rings

Saturn's rings may be very old, dating to the formation of Saturn itself. There are two main theories regarding the origin of Saturn's inner rings. One theory, originally proposed by Édouard Roche in the 19th century, is that the rings were once a moon of Saturn (named Veritas, after a Roman goddess who hid in a well) whose orbit decayed until it came close enough to be ripped apart by tidal forces (see Roche limit). A variation on this theory is that this moon disintegrated after being struck by a large comet or asteroid. The second theory is that the rings were never part of a moon, but are instead left over from the original nebular material from which Saturn formed.Wikipedia:Citation needed

<templatestyles src="Multiple_image/styles.css" />

Saturn's rings

and moons

Tethys, Hyperion and Prometheus

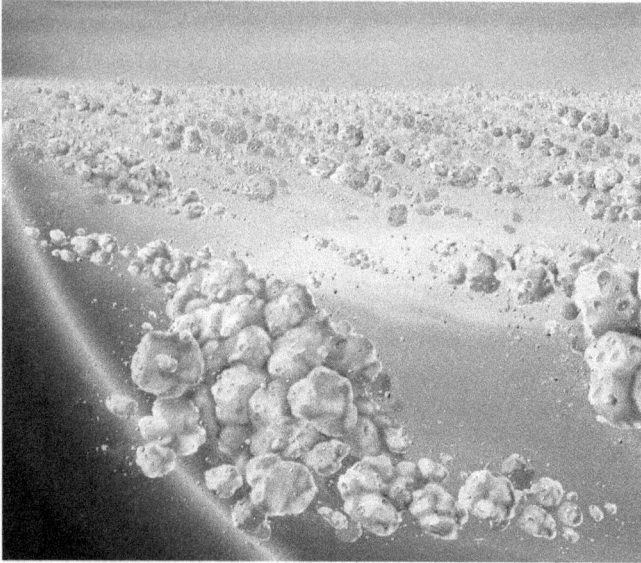

Figure 49: *A 2007 artist impression of the aggregates of icy particles that form the 'solid' portions of Saturn's rings. These elongated clumps are continually forming and dispersing. The largest particles are a few m across.*

Tethys and Janus

A more traditional version of the disrupted-moon theory is that the rings are composed of debris from a moon 400 to 600 km in diameter, slightly larger than Mimas. The last time there were collisions large enough to be likely to disrupt a moon that large was during the Late Heavy Bombardment some four billion years ago.

A more recent variant of this type of theory by R. M. Canup is that the rings could represent part of the remains of the icy mantle of a much larger, Titan-sized, differentiated moon that was stripped of its outer layer as it spiraled into the planet during the formative period when Saturn was still surrounded by a gaseous nebula. This would explain the scarcity of rocky material within the rings. The rings would initially have been much more massive ($\approx 1,000$ times) and broader than at present; material in the outer portions of the rings would have coalesced into the moons of Saturn out to Tethys, also explaining the lack of rocky material in the composition of most of these moons. Subsequent collisional or cryovolcanic evolution of Enceladus might then have caused selective loss of ice from this moon, raising its density to its current value of 1.61 g/cm^3, compared to values of 1.15 for Mimas and 0.97 for Tethys.

The idea of massive early rings was subsequently extended to explain the formation of Saturn's moons out to Rhea. If the initial massive rings contained chunks of rocky material (>100 km across) as well as ice, these silicate bodies would have accreted more ice and been expelled from the rings, due to gravitational interactions with the rings and tidal interaction with Saturn, into progressively wider orbits. Within the Roche limit, bodies of rocky material are dense enough to accrete additional material, whereas less-dense bodies of ice are not. Once outside the rings, the newly formed moons could have continued to evolve through random mergers. This process may explain the variation in silicate content of Saturn's moons out to Rhea, as well as the trend towards less silicate content closer to Saturn. Rhea would then be the oldest of the moons formed from the primordial rings, with moons closer to Saturn being progressively younger.

The brightness and purity of the water ice in Saturn's rings has been cited as evidence that the rings are much younger than Saturn, perhaps just 100 million years old, as the infall of meteoric dust would have led to darkening of the rings. However, new research indicates that the B Ring may be massive enough to have diluted infalling material and thus avoided substantial darkening over the age of the Solar System. Ring material may be recycled as clumps form within the rings and are then disrupted by impacts. This would explain the apparent youth of some of the material within the rings. Further evidence supporting a young ring theory has been gathered by researchers analyzing data from the Cassini Titan Radar Mapper, which focused on analyzing the proportion of rocky silicates contained within the C ring.

The *Cassini* UVIS team, led by Larry Esposito, used stellar occultation to discover 13 objects, ranging from 27 m to 10 km across, within the F ring. They are translucent, suggesting they are temporary aggregates of ice boulders a few m across. Esposito believes this to be the basic structure of the Saturnian rings, particles clumping together, then being blasted apart.

Subdivisions and structures within the rings

The densest parts of the Saturnian ring system are the A and B Rings, which are separated by the Cassini Division (discovered in 1675 by Giovanni Domenico Cassini). Along with the C Ring, which was discovered in 1850 and is similar in character to the Cassini Division, these regions constitute the *main rings*. The main rings are denser and contain larger particles than the tenuous *dusty rings*. The latter include the D Ring, extending inward to Saturn's cloud tops, the G and E Rings and others beyond the main ring system. These diffuse rings are characterised as "dusty" because of the small size of their particles (often about a μm); their chemical composition is, like the main rings, almost entirely water ice. The narrow F Ring, just off the outer edge of the A Ring, is more difficult to categorize; parts of it are very dense, but it also contains a great deal of dust-size particles.

File:Saturn's rings dark side mosaic.jpg

Natural-color mosaic of *Cassini* narrow-angle camera images of the unilluminated side of Saturn's D, C, B, A and F rings (left to right) taken on May 9, 2007.

Physical parameters of the rings

Notes:

[1] Names as designated by the International Astronomical Union, unless otherwise noted. Broader separations between named rings are termed *divisions*, while narrower separations within named rings are called *gaps*.

[2] Data mostly from the Gazetteer of Planetary Nomenclature[214], a NASA factsheet[215] and several papers.

[3] distance is to centre of gaps, rings and ringlets that are narrower than 1,000 km

[4] unofficial name

Major subdivisions

Name[1]	Distance from Saturn's center (km)[2]	Width (km)[2]	Named after
D Ring	66,900 – 74,510	7,500	
C Ring	74,658 – 92,000	17,500	
B Ring	92,000 – 117,580	25,500	
Cassini Division	117,580 – 122,170	4,700	Giovanni Cassini
A ring	122,170 – 136,775	14,600	
Roche Division	136,775 – 139,380	2,600	Édouard Roche
F Ring	140,180 [3]	30 – 500	
Janus/Epimetheus Ring[4]	149,000 – 154,000	5,000	Janus and Epimetheus
G Ring	166,000 – 175,000	9,000	
Methone Ring Arc[4]	194,230	?	Methone
Anthe Ring Arc[4]	197,665	?	Anthe

Pallene Ring[4]	211,000 – 213,500	2,500	Pallene
E Ring	180,000 – 480,000	300,000	
Phoebe Ring	~4,000,000 – >13,000,000		Phoebe

C Ring structures

Name[1]	Distance from Saturn's center (km)[2]	Width (km)[2]	Named after
Colombo Gap	77,870 [3]	150	Giuseppe "Bepi" Colombo
Titan Ringlet	77,870 [3]	25	Titan, moon of Saturn
Maxwell Gap	87,491 [3]	270	James Clerk Maxwell
Maxwell Ringlet	87,491 [3]	64	James Clerk Maxwell
Bond Gap	88,700 [3]	30	William Cranch Bond and George Phillips Bond
1.470R_S Ringlet	88,716 [3]	16	its radius
1.495R_S Ringlet	90,171 [3]	62	its radius
Dawes Gap	90,210 [3]	20	William Rutter Dawes

Cassini Division structures

• Source:

Name[1]	Distance from Saturn's center (km)[2]	Width (km)[2]	Named after
Huygens Gap	117,680 [3]	285–400	Christiaan Huygens
Huygens Ringlet	117,848 [3]	~17	Christiaan Huygens
Herschel Gap	118,234 [3]	102	William Herschel
Russell Gap	118,614 [3]	33	Henry Norris Russell
Jeffreys Gap	118,950 [3]	38	Harold Jeffreys
Kuiper Gap	119,405 [3]	3	Gerard Kuiper
Laplace Gap	119,967 [3]	238	Pierre-Simon Laplace
Bessel Gap	120,241 [3]	10	Friedrich Bessel
Barnard Gap	120,312 [3]	13	Edward Emerson Barnard

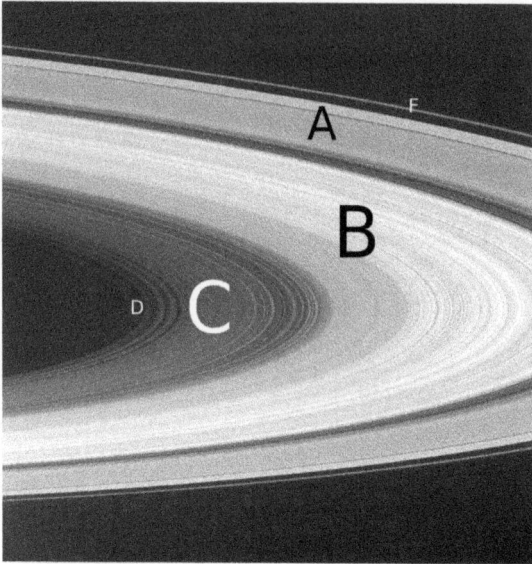

Figure 50: *The illuminated side of Sat-*
urn's rings with the major subdivisions labeled

A Ring structures

Name[1]	Distance from Saturn's center (km)[2]	Width (km)[2]	Named after
Encke Gap	133,589 [3]	325	Johann Encke
Keeler Gap	136,505 [3]	35	James Keeler

Figure 51: *A Cassini image of the faint D Ring, with the inner C Ring below*

File:Saturn's rings in visible light and radio.jpg

Oblique (4 degree angle) *Cassini* images of Saturn's C, B, and A rings (left to right; the F ring is faintly visible in the full size upper image if viewed at sufficient brightness). Upper image: natural color mosaic of *Cassini* narrow-angle camera photos of the illuminated side of the rings taken on December 12, 2004. Lower image: simulated view constructed from a radio occultation observation conducted on May 3, 2005. Color in the lower image is used to represent information about ring particle sizes (see the caption of the article's second image for an explanation).

D Ring

The D Ring is the innermost ring, and is very faint. In 1980, Voyager 1 detected within this ring three ringlets designated D73, D72 and D68, with D68 being the discrete ringlet nearest to Saturn. Some 25 years later, Cassini images showed that D72 had become significantly broader and more diffuse, and had moved planetward by 200 km.

Figure 52: *View of the outer C Ring; the Maxwell Gap with the Maxwell Ringlet on its right side are above and right of center. The Bond Gap is above a broad light band towards the upper right; the Dawes Gap is within a dark band just below the upper right corner.*

Present in the D Ring is a finescale structure with waves 30 km apart. First seen in the gap between the C Ring and D73, the structure was found during Saturn's 2009 equinox to extend a radial distance of 19,000 km from the D Ring to the inner edge of the B Ring. The waves are interpreted as a spiral pattern of vertical corrugations of 2 to 20 m amplitude; the fact that the period of the waves is decreasing over time (from 60 km in 1995 to 30 km by 2006) allows a deduction that the pattern may have originated in late 1983 with the impact of a cloud of debris (with a mass of $\approx 10^{12}$ kg) from a disrupted comet that tilted the rings out of the equatorial plane. A similar spiral pattern in Jupiter's main ring has been attributed to a perturbation caused by impact of material from Comet Shoemaker-Levy 9 in 1994.

C Ring

The C Ring is a wide but faint ring located inward of the B Ring. It was discovered in 1850 by William and George Bond, though William R. Dawes and Johann Galle also saw it independently. William Lassell termed it the

"Crepe Ring" because it seemed to be composed of darker material than the brighter A and B Rings.[216]

Its vertical thickness is estimated at 5 m its mass at around 1.1×10^{18} kg, and its optical depth varies from 0.05 to 0.12.Wikipedia:Citation needed That is, between 5 and 12 percent of light shining perpendicularly through the ring is blocked, so that when seen from above, the ring is close to transparent. The 30-km wavelength spiral corrugations first seen in the D Ring were observed during Saturn's equinox of 2009 to extend throughout the C Ring (see above).

Colombo Gap and Titan Ringlet

The Colombo Gap lies in the inner C Ring. Within the gap lies the bright but narrow Colombo Ringlet, centered at 77,883 km from Saturn's center, which is slightly elliptical rather than circular. This ringlet is also called the Titan Ringlet as it is governed by an orbital resonance with the moon Titan. At this location within the rings, the length of a ring particle's apsidal precession is equal to the length of Titan's orbital motion, so that the outer end of this eccentric ringlet always points towards Titan.

Maxwell Gap and Ringlet

The Maxwell Gap lies within the outer part of the C Ring. It also contains a dense non-circular ringlet, the Maxwell Ringlet. In many respects this ringlet is similar to the ε ring of Uranus. There are wave-like structures in the middle of both rings. While the wave in the ε ring is thought to be caused by Uranian moon Cordelia, no moon has been discovered in the Maxwell gap as of July 2008.

B Ring

The B Ring is the largest, brightest, and most massive of the rings. Its thickness is estimated as 5 to 15 m and its optical depth varies from 0.4 to greater than 5, meaning that >99% of the light passing through some parts of the B Ring is blocked. The B Ring contains a great deal of variation in its density and brightness, nearly all of it unexplained. These are concentric, appearing as narrow ringlets, though the B Ring does not contain any gaps.Wikipedia:Citation needed. In places, the outer edge of the B Ring contains vertical structures deviating up to 2.5 km from the main ring plane.

A 2016 study of spiral density waves using stellar occultations indicated that the B Ring's surface density is in the range of 40 to 140 g/cm^2, lower than previously believed, and that the ring's optical depth has little correlation with its mass density (a finding previously reported for the A and C rings). The total mass of the B Ring was estimated to be somewhere in the range of 7 to 24×10^{18} kg. This compares to a mass for Mimas of 37.5×10^{18} kg.

<templatestyles src="Multiple_image/styles.css" />

High resolution (about 3 km per pixel) color view of the inner-central B Ring (98,600 to 105,500 km from Saturn's center). The structures shown (from 40 km wide ringlets at center to 300-500 km wide bands at right) remain sharply defined at scales below the resolution of the image.

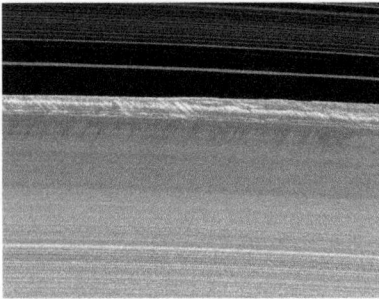

The B Ring's outer edge, viewed near equinox, where shadows are cast by vertical structures up to 2.5 km high, probably created by unseen embedded moonlets. The Cassini Division is at top.

Spokes

Until 1980, the structure of the rings of Saturn was explained as being caused exclusively by the action of gravitational forces. Then images from the Voyager spacecraft showed radial features in the B Ring, known as *spokes*, which could not be explained in this manner, as their persistence and rotation around the rings was not consistent with gravitational orbital mechanics. The spokes appear dark in backscattered light, and bright in forward-scattered light (see images in Gallery); the transition occurs at a phase angle near 60°. The leading theory regarding the spokes' composition is that they consist of microscopic dust particles suspended away from the main ring by electrostatic repulsion, as

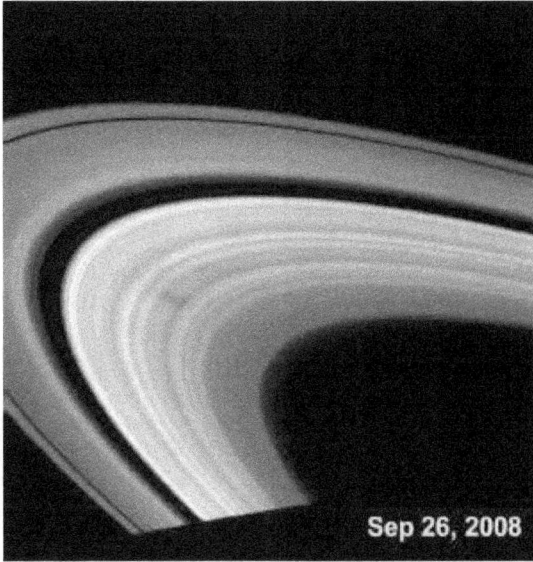

Figure 53: *Dark spokes mark the B ring's sunlit side in low phase angle Cassini images. This is a low-bitrate video. Full-size video with high bitrate of 471 kbit/s; GIF version (400 × 400 pixels, file size: 2.21 MB)*

they rotate almost synchronously with the magnetosphere of Saturn. The precise mechanism generating the spokes is still unknown, although it has been suggested that the electrical disturbances might be caused by either lightning bolts in Saturn's atmosphere or micrometeoroid impacts on the rings.

The spokes were not observed again until some twenty-five years later, this time by the Cassini space probe. The spokes were not visible when Cassini arrived at Saturn in early 2004. Some scientists speculated that the spokes would not be visible again until 2007, based on models attempting to describe their formation. Nevertheless, the Cassini imaging team kept looking for spokes in images of the rings, and they were next seen in images taken on 5 September 2005.

The spokes appear to be a seasonal phenomenon, disappearing in the Saturnian midwinter and midsummer and reappearing as Saturn comes closer to equinox. Suggestions that the spokes may be a seasonal effect, varying with Saturn's 29.7-year orbit, were supported by their gradual reappearance in the later years of the Cassini mission.

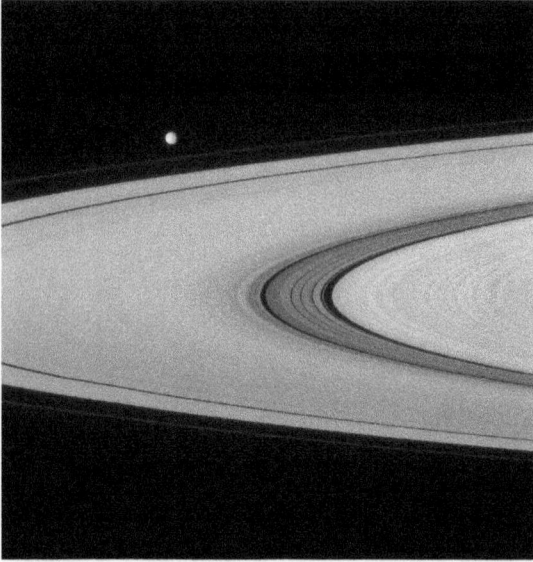

Figure 54: *The Cassini Division imaged from the Cassini space-craft. The Huygens Gap lies at its right border; the Laplace Gap is to-wards the center. A number of other, narrower gaps are also present.*

Moonlet

In 2009, during equinox, a moonlet embedded in the B ring was discovered from the shadow it cast. It is estimated to be 400 m (1,300 ft) in diameter. The moonlet was given the provisional designation S/2009 S 1.

Cassini Division

The Cassini Division is a region 4,800 km (3,000 mi) in width between Saturn's A ring and B Ring. It was discovered in 1675 by Giovanni Cassini at the Paris Observatory using a refracting telescope that had a 2.5-inch objective lens with a 20-foot-long focal length and a 90x magnification.[217] From Earth it appears as a thin black gap in the rings. However, *Voyager* discovered that the gap is itself populated by ring material bearing much similarity to the C Ring. The division may appear bright in views of the unlit side of the rings, since the relatively low density of material allows more light to be transmitted through the thickness of the rings (see second image in gallery).Wikipedia:Citation needed

Figure 55: *The central ringlet of the A Ring's Encke Gap coincides with Pan's orbit, implying its particles oscillate in horseshoe orbits.*

The inner edge of the Cassini Division is governed by a strong orbital resonance. Ring particles at this location orbit twice for every orbit of the moon Mimas. The resonance causes Mimas' pulls on these ring particles to accumulate, destabilizing their orbits and leading to a sharp cutoff in ring density. Many of the other gaps between ringlets within the Cassini Division, however, are unexplained.Wikipedia:Citation needed

Huygens Gap

The Huygens Gap is located at the inner edge of the Cassini Division. It contains the dense, eccentric Huygens Ringlet in the middle. This ringlet exhibits irregular azimuthal variations of geometrical width and optical depth, which may be caused by the nearby 2:1 resonance with Mimas and the influence of the eccentric outer edge of the B-ring. There is an additional narrow ringlet just outside the Huygens Ringlet.

A Ring

The A Ring is the outermost of the large, bright rings. Its inner boundary is the Cassini Division and its sharp outer boundary is close to the orbit of the small moon Atlas. The A Ring is interrupted at a location 22% of the ring width from its outer edge by the Encke Gap. A narrower gap 2% of the ring width from the outer edge is called the Keeler Gap.

The thickness of the A Ring is estimated to be 10 to 30 m, its surface density from 35 to 40 g/cm^2 and its total mass as 4 to 5×10^{18} kg (just under the mass of Hyperion). Its optical depth varies from 0.4 to 0.9.

Similarly to the B Ring, the A Ring's outer edge is maintained by an orbital resonance, in this case the 7:6 resonance with Janus and Epimetheus.Wikipedia:Citation needed Other orbital resonances also excite many spiral density waves in the A Ring (and, to a lesser extent, other rings as well), which account for most of its structure. These waves are described by the same physics that describes the spiral arms of galaxies. Spiral bending waves, also present in the A Ring and also described by the same theory, are vertical corrugations in the ring rather than compression waves.Wikipedia:Citation needed

In April 2014, NASA scientists reported observing the possible formative stage of a new moon near the outer edge of the A Ring.

Encke Gap

The Encke Gap is a 325-km-wide gap within the A ring, centered at a distance of 133,590 km from Saturn's center. It is caused by the presence of the small moon Pan, which orbits within it. Images from the *Cassini* probe have shown that there are at least three thin, knotted ringlets within the gap. Spiral density waves visible on both sides of it are induced by resonances with nearby moons exterior to the rings, while Pan induces an additional set of spiraling wakes (see image in gallery).

Johann Encke himself did not observe this gap; it was named in honour of his ring observations. The gap itself was discovered by James Edward Keeler in 1888. The second major gap in the A ring, discovered by *Voyager*, was named the Keeler Gap in his honor.

The Encke Gap is a *gap* because it is entirely within the A Ring. There was some ambiguity between the terms *gap* and *division* until the IAU clarified the definitions in 2008; before that, the separation was sometimes called the "Encke Division".

Figure 56: *Waves in the Keeler gap edges induced by the orbital motion of Daphnis (see also a stretched closeup view in the gallery).*

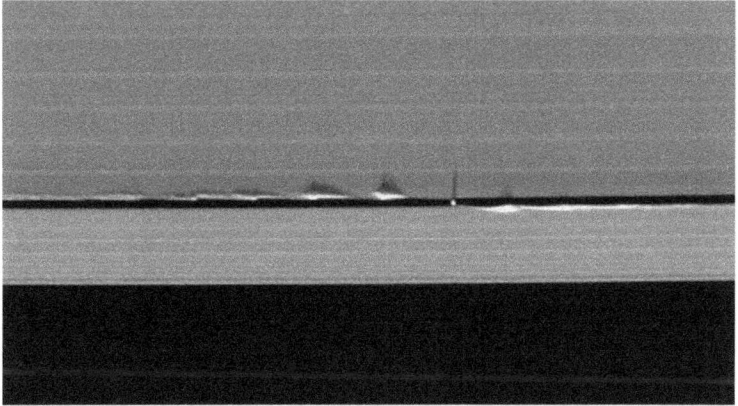

Figure 57: *Near Saturn's equinox, Daphnis and its waves cast shadows on the A Ring.*

Keeler Gap

The Keeler Gap is a 42-km-wide gap in the A ring, approximately 250 km from the ring's outer edge. The small moon Daphnis, discovered 1 May 2005, orbits within it, keeping it clear. The moon's passage induces waves in the edges of the gap (this is also influenced by its slight orbital eccentricity). Because the orbit of Daphnis is slightly inclined to the ring plane, the waves have a component that is perpendicular to the ring plane, reaching a distance of 1500 m "above" the plane.

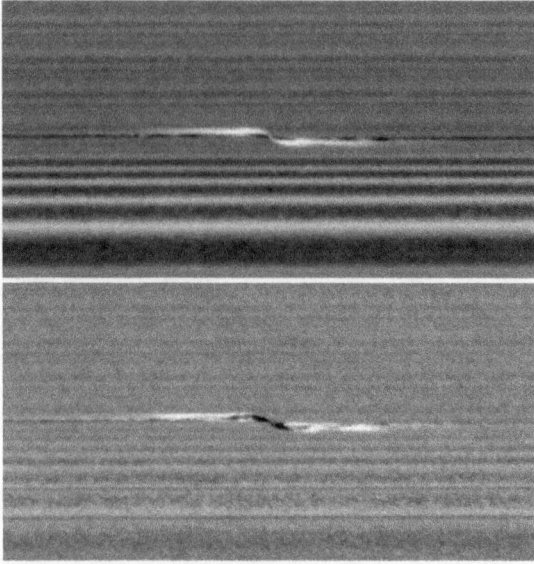

Figure 58: *Propeller moonlet Santos-Dumont from lit (top) and unlit sides of rings*

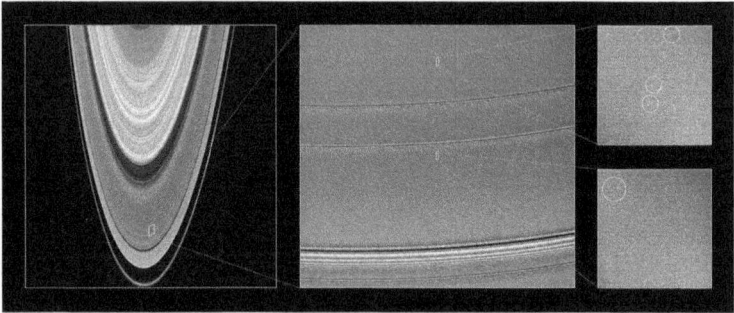

Figure 59: *Location of the first four moonlets detected in the A ring.*

The Keeler gap was discovered by *Voyager*, and named in honor of the astronomer James Edward Keeler. Keeler had in turn discovered and named the Encke Gap in honor of Johann Encke.

Propeller moonlets

In 2006, four tiny "moonlets" were found in *Cassini* images of the A Ring. The moonlets themselves are only about a hundred m in diameter, too small to be seen directly; what *Cassini* sees are the "propeller"-shaped disturbances

Figure 60: *The Roche Division (passing through image center) between the A Ring and the narrow F Ring. Atlas can be seen within it. The Encke and Keeler gaps are also visible.*

the moonlets create, which are several km across. It is estimated that the A Ring contains thousands of such objects. In 2007, the discovery of eight more moonlets revealed that they are largely confined to a 3,000 km belt, about 130,000 km from Saturn's center, and by 2008 over 150 propeller moonlets had been detected. One that has been tracked for several years has been nicknamed *Bleriot*.

Roche Division

The separation between the A ring and the F Ring has been named the Roche Division in honor of the French physicist Édouard Roche. The Roche Division should not be confused with the Roche limit which is the distance at which a large object is so close to a planet (such as Saturn) that the planet's tidal forces will pull it apart. Lying at the outer edge of the main ring system, the Roche Division is in fact close to Saturn's Roche limit, which is why the rings have been unable to accrete into a moon.

Like the Cassini Division, the Roche Division is not empty but contains a sheet of material.Wikipedia:Citation needed The character of this material is similar to the tenuous and dusty D, E, and G Rings.Wikipedia:Citation needed Two

Figure 61: *The small moons Pandora (left) and Prometheus (right) orbit on either side of the F ring. Prometheus acts as a ring shepherd and is followed by dark channels that it has carved*[220] *into the inner strands of the ring.*

locations in the Roche Division have a higher concentration of dust than the rest of the region. These were discovered by the *Cassini* probe imaging team and were given temporary designations: R/2004 S 1, which lies along the orbit of the moon Atlas; and R/2004 S 2, centered at 138,900 km from Saturn's center, inward of the orbit of Prometheus.[218,219]

F Ring

The F Ring is the outermost discrete ring of Saturn and perhaps the most active ring in the Solar System, with features changing on a timescale of hours. It is located 3,000 km beyond the outer edge of the A ring. The ring was discovered in 1979 by the Pioneer 11 imaging team. It is very thin, just a few hundred km in radial extent. While the traditional view has been that it is held together by two shepherd moons, Prometheus and Pandora, which orbit inside and outside it, recent studies indicate that only Prometheus contributes to the confinement. Numerical simulations suggest the ring was formed when Prometheus and Pandora collided with each other and were partially disrupted.

Recent closeup images from the Cassini probe show that the F Ring consists of one core ring and a spiral strand around it. They also show that when

Figure 62: *The outer rings seen back-illuminated by the Sun*

Prometheus encounters the ring at its apoapsis, its gravitational attraction creates kinks and knots in the F Ring as the moon 'steals' material from it, leaving a dark channel in the inner part of the ring (see video link and additional F Ring images in gallery). Since Prometheus orbits Saturn more rapidly than the material in the F ring, each new channel is carved about 3.2 degrees in front of the previous one.

In 2008, further dynamism was detected, suggesting that small unseen moons orbiting within the F Ring are continually passing through its narrow core because of perturbations from Prometheus. One of the small moons was tentatively identified as S/2004 S 6.

File:F Ring perturbations PIA08412.jpg

A mosaic of 107 images showing 255° (about 70%) of the F Ring as it would appear if straightened out, showing the kinked primary strand and the spiral secondary strand. The radial width (top to bottom) is 1,500 km.

Outer rings

Janus/Epimetheus Ring

A faint dust ring is present around the region occupied by the orbits of Janus and Epimetheus, as revealed by images taken in forward-scattered light by the Cassini spacecraft in 2006. The ring has a radial extent of about 5,000 km.[221] Its source is particles blasted off the moons' surfaces by meteoroid impacts, which then form a diffuse ring around their orbital paths.

G Ring

The G Ring (see last image in gallery) is a very thin, faint ring about halfway between the F Ring and the beginning of the E Ring, with its inner edge about 15,000 km inside the orbit of Mimas. It contains a single distinctly brighter arc near its inner edge (similar to the arcs in the rings of Neptune) that extends about one sixth of its circumference, centered on the half-km diameter moonlet Aegaeon, which is held in place by a 7:6 orbital resonance with Mimas. The arc is believed to be composed of icy particles up to a few m in diameter, with the rest of the G Ring consisting of dust released from within the arc. The radial width of the arc is about 250 km, compared to a width of 9,000 km for the G Ring as a whole. The arc is thought to contain matter equivalent to a small icy moonlet about a hundred m in diameter. Dust released from Aegaeon and other source bodies within the arc by micrometeoroid impacts drifts outward from the arc because of interaction with Saturn's magnetosphere (whose plasma corotates with Saturn's magnetic field, which rotates much more rapidly than the orbital motion of the G Ring). These tiny particles are steadily eroded away by further impacts and dispersed by plasma drag. Over the course of thousands of years the ring gradually loses mass, which is replenished by further impacts on Aegaeon.

Methone Ring Arc

A faint ring arc, first detected in September 2006, covering a longitudinal extent of about 10 degrees is associated with the moon Methone. The material in the arc is believed to represent dust ejected from Methone by micrometeoroid impacts. The confinement of the dust within the arc is attributable to a 14:15 resonance with Mimas (similar to the mechanism of confinement of the arc within the G ring). Under the influence of the same resonance, Methone librates back and forth in its orbit with an amplitude of 5° of longitude.

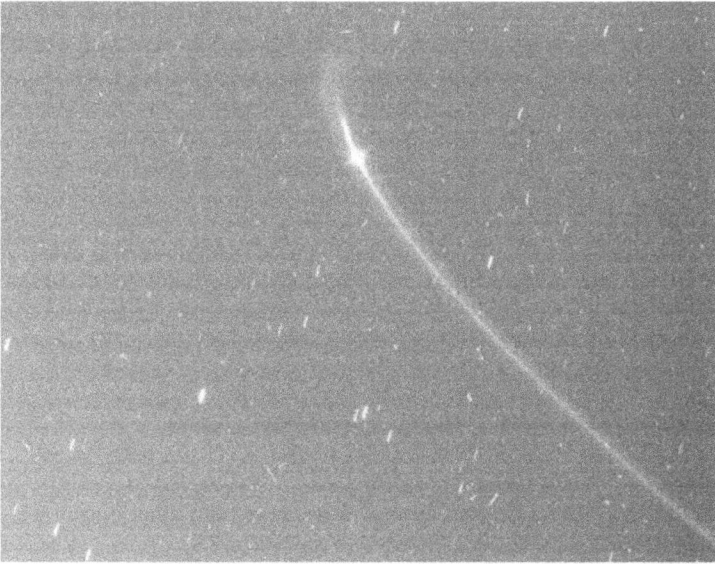

Figure 63: *The Anthe Ring Arc - the bright spot is Anthe*

Anthe Ring Arc

A faint ring arc, first detected in June 2007, covering a longitudinal extent of about 20 degrees is associated with the moon Anthe. The material in the arc is believed to represent dust knocked off Anthe by micrometeoroid impacts. The confinement of the dust within the arc is attributable to a 10:11 resonance with Mimas. Under the influence of the same resonance, Anthe drifts back and forth in its orbit over 14° of longitude.

Pallene Ring

A faint dust ring shares Pallene's orbit, as revealed by images taken in forward-scattered light by the *Cassini* spacecraft in 2006. The ring has a radial extent of about 2,500 km. Its source is particles blasted off Pallene's surface by meteoroid impacts, which then form a diffuse ring around its orbital path.

E Ring

The E Ring is the second outermost ring and is extremely wide; it consists of many tiny (micron and sub-micron) particles of water ice with silicates, carbon dioxide and ammonia. The E Ring is distributed between the orbits of Mimas and Titan. Unlike the other rings, it is composed of microscopic particles rather than macroscopic ice chunks. In 2005, the source of the E Ring's material was determined to be cryovolcanic plumes emanating from the "tiger stripes" of the south polar region of the moon Enceladus. Unlike the main rings, the E Ring is more than 2,000 km thick and increases with its distance from Enceladus. Tendril-like structures observed within the E Ring can be related to the emissions of the most active south polar jets of Enceladus.

Particles of the E Ring tend to accumulate on moons that orbit within it. The equator of the leading hemisphere of Tethys is tinted slightly blue due to in-falling material.[222] The trojan moons Telesto, Calypso, Helene and Polydeuces are particularly affected as their orbits move up and down the ring plane. This results in their surfaces being coated with bright material that smooths out features.[223] <templatestyles src="Multiple_image/styles.css" />

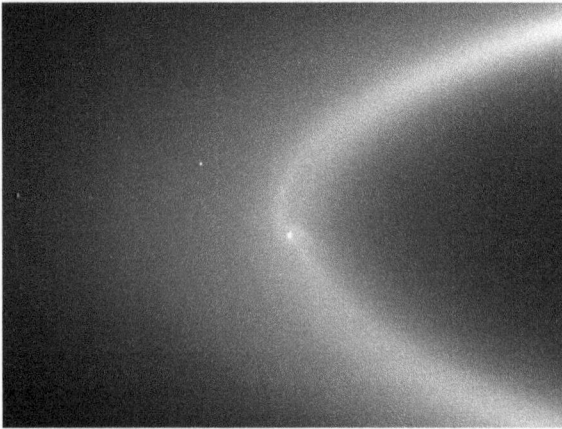

The backlit E ring, with Enceladus silhouetted against it. The moon's south polar jets erupt brightly below it.

Close-up of the south polar geysers of Enceladus, the source of the E Ring.
<templatestyles src="Multiple_image/styles.css" />

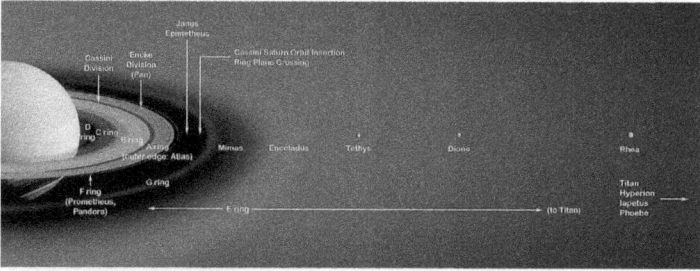

Side view of Saturn system, showing Enceladus in relation to the E Ring

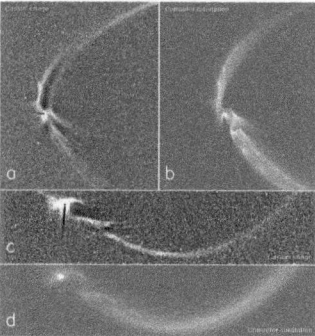

E Ring tendrils from Enceladus geysers - comparison of images (a, c) with computer simulations

Phoebe ring

In October 2009, the discovery of a tenuous disk of material just interior to the orbit of Phoebe was reported. The disk was aligned edge-on to Earth at the time of discovery. This disk can be loosely described as another ring. Although very large (as seen from Earth, the apparent size of two full moons), the ring is virtually invisible. It was discovered using NASA's infrared Spitzer Space Telescope,[224] and was seen over the entire range of the observations, which extended from 128 to 207 times the radius of Saturn, with calculations indicating that it may extend outward up to 300 Saturn radii and inward to the orbit of Iapetus at 59 Saturn radii. The ring was subsequently studied using the WISE, *Herschel* and *Cassini* spacecraft; WISE observations show that it extends from at least between 50 and 100 to 270 Saturn radii (the inner edge is lost in the planet's glare). Data obtained with WISE indicate the ring particles are small; those with radii of greater than 10 cm comprise 10% or less of the cross-sectional area.

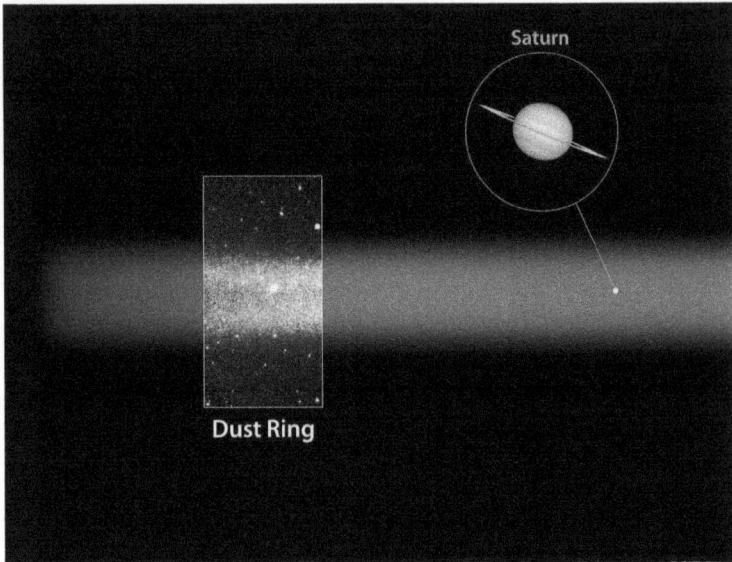

Figure 64: *The Phoebe ring's huge extent dwarfs the
main rings. Inset: 24 μm Spitzer image of part of the ring*

Phoebe orbits the planet at a distance ranging from 180 to 250 radii. The ring
has a thickness of about 40 radii. Because the ring's particles are presumed
to have originated from impacts (micrometeoroid and larger) on Phoebe, they
should share its retrograde orbit, which is opposite to the orbital motion of
the next inner moon, Iapetus. This ring lies in the plane of Saturn's orbit, or
roughly the ecliptic, and thus is tilted 27 degrees from Saturn's equatorial plane
and the other rings. Phoebe is inclined by 5° with respect to Saturn's orbit
plane (often written as 175°, due to Phoebe's retrograde orbital motion), and
its resulting vertical excursions above and below the ring plane agree closely
with the ring's observed thickness of 40 Saturn radii.

The existence of the ring was proposed in the 1970s by Steven Soter. The
discovery was made by Anne J. Verbiscer and Michael F. Skrutskie (of the
University of Virginia) and Douglas P. Hamilton (of the University of Maryland, College Park). The three had studied together at Cornell University as
graduate students.

Ring material migrates inward due to reemission of solar radiation, with a
speed inversely proportional to particle size; a 3 cm particle would migrate
from the vicinity of Phoebe to that of Iapetus over the age of the Solar System. The material would thus strike the leading hemisphere of Iapetus. Infall

of this material causes a slight darkening and reddening of the leading hemisphere of Iapetus (similar to what is seen on the Uranian moons Oberon and Titania) but does not directly create the dramatic two-tone coloration of that moon. Rather, the infalling material initiates a positive feedback thermal self-segregation process of ice sublimation from warmer regions, followed by vapor condensation onto cooler regions. This leaves a dark residue of "lag" material covering most of the equatorial region of Iapetus's leading hemisphere, which contrasts with the bright ice deposits covering the polar regions and most of the trailing hemisphere.

Possible ring system around Rhea

Saturn's second largest moon Rhea has been hypothesized to have a tenuous ring system of its own consisting of three narrow bands embedded in a disk of solid particles. These putative rings have not been imaged, but their existence has been inferred from *Cassini* observations in November 2005 of a depletion of energetic electrons in Saturn's magnetosphere near Rhea. The Magnetospheric Imaging Instrument (MIMI) observed a gentle gradient punctuated by three sharp drops in plasma flow on each side of the moon in a nearly symmetric pattern. This could be explained if they were absorbed by solid material in the form of an equatorial disk containing denser rings or arcs, with particles perhaps several dm to approximately a m in diameter. A more recent piece of evidence consistent with the presence of Rhean rings is a set of small ultraviolet-bright spots distributed in a line that extends three quarters of the way around the moon's circumference, within 2 degrees of the equator. The spots have been interpreted as the impact points of deorbiting ring material. However, targeted observations by *Cassini* of the putative ring plane from several angles have turned up nothing, suggesting that another explanation for these enigmatic features is needed.

Gallery

Figure 65: *Saturn, draped with the rings' shadows, behind the rings*

Figure 66: *Cassini image mosaic of the unlit side of the outer C Ring (bottom) and inner B Ring (top) near Saturn's equinox, showing multiple views of the shadow of Mimas. The shadow is attenuated by the denser B ring. The Maxwell Gap is below center.*

Figure 67: *A spiral density wave in Saturn's inner B Ring which forms at a 2:1 orbital resonance with Janus. The wavelength decreases as the wave propagates away from the resonance, so the apparent foreshortening in the image is illusory.*

Figure 68: *Natural color view of the outer C Ring and B Ring.*

Figure 69: *Dark B Ring spokes in a low-phase-angle Cassini image of the rings' unlit side. Left of center, two dark gaps (the larger being the Huygens Gap) and the bright (from this viewing geometry) ringlets to their left comprise the Cassini Division.*

Figure 70: *Cassini image of the sun-lit side of the rings taken in 2009 at a phase angle of 144°, with bright B Ring spokes.*

Figure 71: *Pan's motion through the A ring's Encke Gap induces edge waves and (non-self-propagating) spiraling wakes ahead of and inward of it[225]. The other more tightly wound bands are spiral density waves.*

Figure 72: *Radially stretched (4x) view of the Keeler Gap edge waves induced by Daphnis.*

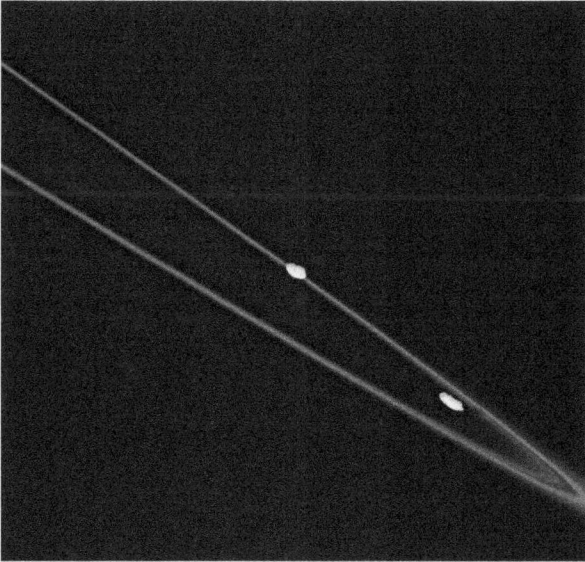

Figure 73: *Prometheus (at center) and Pandora orbit just inside and outside the F Ring, but only Prometheus acts as a ring shepherd.*

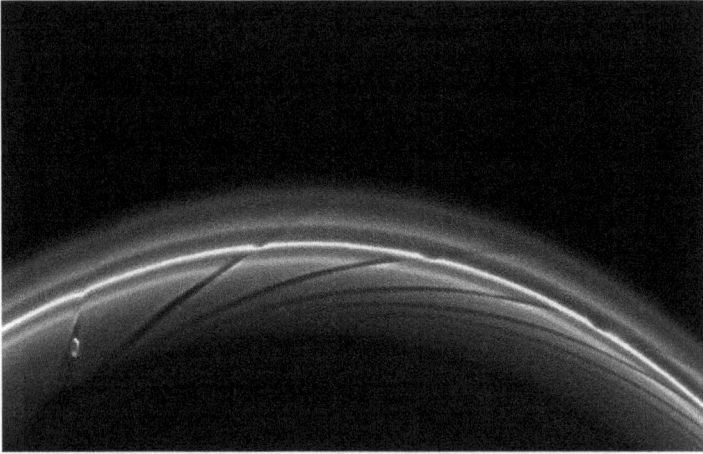

Figure 74: *Prometheus near apoapsis carving a dark channel in the F Ring (with older channels to the right). A movie of the process may be viewed at the Cassini Imaging Team website or YouTube.*

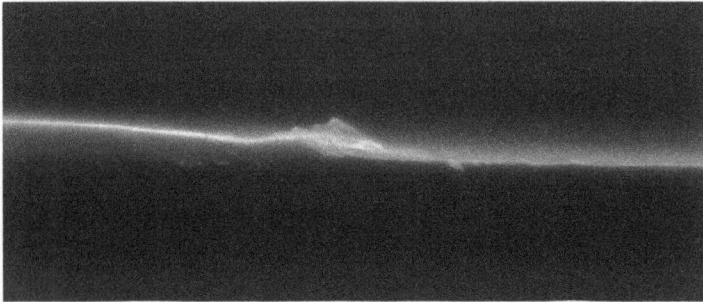

Figure 75: *F ring dynamism, probably due to perturbing effects of small moonlets orbiting close to or through the ring's core.*

Figure 76: *Saturn's shadow truncates the backlit G Ring and its bright inner arc. A video showing the arc's orbital motion may be viewed on YouTube or the Cassini Imaging Team website.*

External links

 Wikimedia Commons has media related to *Rings of Saturn*.

 Wikimedia Commons has media related to *Saturn (Rings)*.

- Planetary Rings Node: Saturn's Ring System[226]
- Saturn's Rings[227] by NASA's Solar System Exploration[228]
- Rings of Saturn nomenclature[229] from the USGS planetary nomenclature page[230]
- Biggest Ring Around Saturn Just Got Supersized[231] (retrieved 2017-12-20 from Space.com)
- Everything a Curious Mind Should Know About Planetary Ring Systems with Dr Mark Showalter[232] (Waseem Akhtar podcast with Mark Showalter)
- High-resolution animation[233] by Seán Doran of the backlit rings

History of observation and exploration

Saturn (mythology)

Saturn	
Titan of Capitol, wealth, agriculture, liberation, and time	
2nd-century AD Roman bas-relief depicting the god Saturn holding a scythe	
Symbol	Sickle, scythe
Temple	Temple of Saturn
Festivals	The Saturnalia
Personal information	
Consort	Ops
Children	Jupiter, Neptune, Pluto, Juno, Ceres and Vesta
Parents	Caelus and Terra
Siblings	Janus, Ops
Greek equivalent	Cronus

Saturn (Latin: *Saturnus* pronounced [sa'tʊr.nʊs]) is a god in ancient Roman religion, and a character in myth as a god of generation, dissolution, plenty, wealth, agriculture, periodic renewal and liberation. In later developments, he also came to be a god of time. His reign was depicted as a Golden Age of plenty and peace. The Temple of Saturn in the Roman Forum housed the state treasury. In December, he was celebrated at what is perhaps the most famous of the Roman festivals, the Saturnalia, a time of feasting, role reversals, free speech, gift-giving and revelry. Saturn the planet and Saturday are both named after the god.

Mythology

The Roman land preserved the remembrance of a very remote time during which Saturn and Janus reigned on the site of the city before its foundation: the Capitol was called "mons Saturnius.[234] The Romans identified Saturn with the Greek Cronus, whose myths were adapted for Latin literature and Roman art. In particular, Cronus's role in the genealogy of the Greek gods was transferred to Saturn. As early as Livius Andronicus (3rd century BC), Jupiter was called the son of Saturn.[235]

Saturn had two mistresses who represented different aspects of the god. The name of his wife Ops, the Roman equivalent of Greek Rhea, means "wealth, abundance, resources."[236] The association with Ops is considered a later development, however, as this goddess was originally paired with Consus.[237] Earlier was Saturn's association with Lua ("destruction, dissolution, loosening"), a goddess who received the bloodied weapons of enemies destroyed in war.[238]

Under Saturn's rule, humans enjoyed the spontaneous bounty of the earth without labour in the "Golden Age" described by Hesiod and Ovid.

Etymology and epithets

<templatestyles src="Template:Quote_box/styles.css" />

By Saturn they seek to represent that power which maintains the cyclic course of times and seasons. This is the sense that the Greek name of that god bears, for he is called Kronos, which is the same as Chronos or Time. Saturn for his part got his name because he was "sated" with years; the story that he regularly devoured his own children is explained by the fact that time devours the courses of the seasons, and gorges itself "insatiably" on the years that are past. Saturn was enchained by Jupiter to ensure that his circuits did not get out of control, and to constrain him with the bonds of the stars.

—

Quintus Lucilius Balbus as recorded by Marcus Tullius Cicero and translated by P.G. Walsh, *De Natura Deorum (On the Nature of the Gods)*, Book II, Part ii, Section c

According to Varro,[239] Saturn's name was derived from *satu*, meaning "sowing". Even though this etymology looks implausible on linguistic grounds (for the long quantity of the *a* in *Sāturnus* and also because of the epigraphically attested form *Saeturnus*)[240] nevertheless it does reflect an original feature of the god.[241] A more probable etymology connects the name with Etruscan god *Satre* and placenames such as *Satria*, an ancient town of Latium, and *Saturae palus*, a marsh also in Latium. This root may be related to Latin phytonym *satureia*.[242]

Another epithet, variably *Sterculius*, *Sterc utus*, and *Sterces*, referred to his agricultural functions;[243] this derives from *stercus*, "dung" or "manure", referring to re-emergence from death to life.[244] Agriculture was important to Roman identity, and Saturn was a part of archaic Roman religion and ethnic identity. His name appears in the ancient hymn of the Salian priests,[245] and his temple was the oldest known to have been recorded by the pontiffs.

Quintus Lucilius Balbus gives a separate etymology in Cicero's De Natura Deorum (On the Nature of the Gods). In this interpretation, the agricultural aspect of Saturn would be secondary to his primary relation with time and seasons. Since Time consumes all things, Balbus asserts that the name Saturn comes from the Latin word *satis*; Saturn being an anthropomorphic representation of Time, which is filled, or satiated, by all things or all generations. Since agriculture is so closely linked to seasons and therefore an understanding of the cyclical passage of time, it follows that agriculture would then be associated with the deity Saturn.

Temple

The temple of Saturn was located at the base of the Capitoline Hill, according to a tradition recorded by Varro[246] formerly known as *Saturnius Mons*, and a row of columns from the last rebuilding of the temple still stands.[247] The temple was consecrated in 497 BC but the *area Saturni* was built by king Tullus Hostilius as confirmed by archaeological studies conducted by E. Gjerstad.[248] It housed the state treasury (aerarium) throughout Roman history.

Figure 77: *Ruins of the Temple of Saturn (eight columns to the far right) in February 2010, with three columns from the Temple of Vespasian and Titus (left) and the Arch of Septimius Severus (center)*

Festival's time

The position of Saturn's festival in the Roman calendar led to his association with concepts of time, especially the temporal transition of the New Year. In the Greek tradition, Cronus was sometimes conflated with Chronus, "Time," and his devouring of his children taken as an allegory for the passing of generations. The sickle or scythe of Father Time is a remnant of the agricultural implement of Cronus-Saturn, and his aged appearance represents the waning of the old year with the birth of the new, in antiquity sometimes embodied by Aion. In late antiquity, Saturn is syncretized with a number of deities, and begins to be depicted as winged, as is Kairos, "Timing, Right Time".[249]

In Roman religion

Theology and worship

The figure of Saturn is one of the most complex in Roman religion. G. Dumézil refrained from discussing Saturn in his work on Roman religion on the grounds of insufficient knowledge.[250] On the contrary, his follower Dominique Briquel has attempted a thorough interpretation of Saturn utilising Dumézil's three-functional theory of Indoeuropean religion, taking the ancient testimonies and the works of A. Brelich and G. Piccaluga as his basis.[251]

The main difficulty scholars find in studying Saturn is in assessing what is original of his figure and what is due to later hellenising influences. Moreover, some features of the god may be common to Cronus but are nonetheless very ancient and can be considered proper to the Roman god, whereas others are certainly later and arrived after 217 BC, the year in which the Greek customs of the Kronia were introduced into the Saturnalia.[252]

Among the features which are definitely authentic of the Roman god, Briquel identifies:

1. the time of his festival in the calendar, which corresponds to the date of the consecration of his temple (the Greek Cronia on the other hand took place in June–July);
2. his association with *Lua Mater*, and
3. the location of his cult on the Capitol, which goes back to remote times.[253]

These three elements in Briquel's view indicate that Saturn is a sovereign god. The god's strict relationship with the cults of the Capitoline Hill and in particular with Jupiter are highlighted by the legends concerning the refusal of gods Iuventas and Terminus to leave their abode in the shrines on the Capitol when the temple of Jupiter was to be built. These two deities correspond to the helper gods of the sovereign in Vedic religion (Briquel refers to Dhritarashtra and Vidura, the figures of the Mahabharata) and to the Cyclopes and Hecatonchires in Hesiod. Whereas the helper gods belong to the second divine generation they become active only at the level of the third in each of the three instances of India, Greece and Rome, where they become a sort of continuation of Jupiter.[254])

Dumézil postulated a split of the figure of the sovereign god in Indoeuropean religion, which is embodied by Vedic gods Varuna and Mitra.[255] Of the two, the first one shows the aspect of the magic, uncanny, awe inspiring power of creation and destruction, while the second shows the reassuring aspect of guarantor of the legal order in organised social life. Whereas in Jupiter these double features have coalesced, Briquel sees Saturn as showing the characters of a sovereign god of the Varunian type. His nature becomes evident in his mastership over the annual time of crisis around the winter solstice, epitomised in the power of subverting normal codified social order and its rules, which is apparent in the festival of the Saturnalia, in the mastership of annual fertility and renewal, in the power of annihilation present in his paredra Lua, in the fact that he is the god of a timeless era of plenty and bounty before time, which he reinstates at the time of the yearly crisis of the winter solstice. Also, in Roman and Etruscan reckoning Saturn is a wielder of lightning; no other agricultural god (in the sense of specialized human activity) is one.[256] Hence the mastership he has on agriculture and wealth cannot be that of a god of the third function, i.e. of production, wealth, and pleasure, but it stems from his magical lordship over creation and destruction. Although these features are to be found in Greek god Cronus as well, it appears that those features were proper to Roman Saturn's most ancient aspects, such as his presence on the Capitol and his association with Jupiter, who in the stories of the arrival of the Pelasgians in the land of the Sicels[257] and that of the Argei orders human sacrifices to him.[258]

Sacrifices to Saturn were performed according to "Greek rite" (*ritus graecus*), with the head uncovered, in contrast to those of other major Roman deities, which were performed *capite velato*, "with the head covered." Saturn himself, however, was represented as veiled (*involutus*), as for example in a wall painting from Pompeii that shows him holding a sickle and covered with a white veil. This feature is in complete accord with the character of a sovereign god of the Varunian type and is common with German god Odin. Briquel remarks Servius had already seen that the choice of the Greek rite was due to the fact that the god himself is imagined and represented as veiled, thence his sacrifice cannot be carried out by a veiled man: this is an instance of the reversal of the current order of things typical of the nature of the deity as appears in its festival.[259] Plutarch writes his figure is veiled because he is the father of truth.[260]

Pliny notes that the cult statue of Saturn was filled with oil; the exact meaning of this is unclear.[261] Its feet were bound with wool, which was removed only during the Saturnalia.[262] The fact that the statue was filled with oil and the feet were bound with wool may relate back to the myth of "The Castration of Uranus". In this myth Rhea gives Cronus a rock to eat in Zeus' stead, thus tricking Cronus. Although mastership of knots is a feature of Greek origin it is also typical of the Varunian sovereign figure, as apparent e.g. in Odin. Once Zeus was victorious over Cronus, he sets this stone up at Delphi and constantly it is anointed with oil and strands of unwoven wool are placed on it.[263] It wore a red cloak,[264] and was brought out of the temple to take part in ritual processions[265] and *lectisternia*, banquets at which images of the gods were arranged as guests on couches. All these ceremonial details identify a sovereign figure. Briquel concludes that Saturn was a sovereign god of a time that the Romans perceived as no longer actual, that of the legendary origins of the world, before civilization.[266]

Little evidence exists in Italy for the cult of Saturn outside Rome, but his name resembles that of the Etruscan god Satres.[267] The potential cruelty of Saturn was enhanced by his identification with Cronus, known for devouring his own children. He was thus used in translation when referring to gods from other cultures the Romans perceived as severe; he was equated with the Carthaginian god Ba'al Hammon, to whom children were sacrificed, and to Yahweh, whose Sabbath was referred to as *Saturni dies*, "Saturn's day," in a poem by Tibullus, who wrote during the reign of Augustus; eventually this gave rise to the word "Saturday" in English.[268] The identification with Ba'al Hammon later gave rise to the African Saturn, a cult that enjoyed great popularity until the 4th century. It had a popular but also a mysteric character and required child sacrifices. It is also considered as inclining to monotheism.[269] In the ceremony of initiation the myste *intrat sub iugum*, ritual that Leglay compares to the Roman *tigillum sororium*.[270] Even though their origin and theology are completely different the Italic and the African god are both sovereign and master over time and death, fact that has permitted their encounter. Moreover, here Saturn is not the real Italic god but his Greek counterpart Cronus.

Saturnalia

Saturn is associated with a major religious festival in the Roman calendar, *Saturnalia*. Saturnalia celebrated the harvest and sowing, and ran from December 17–23. During Saturnalia, the social restrictions of Rome were relaxed. The figure of Saturn, kept during the year with its legs bound in wool, was released from its bindings for the period of the festival. The revelries of Saturnalia were supposed to reflect the conditions of the lost "Golden Age" before the rule of Saturn was overthrown, not all of them desirable except as a temporary release from civilized constraint. The Greek equivalent was the Kronia.[271]

Macrobius (5th century AD) presents an interpretation of the Saturnalia as a festival of light leading to the winter solstice.[272] The renewal of light and the coming of the new year was celebrated in the later Roman Empire at the *Dies Natalis* of Sol Invictus, the "Birthday of the Unconquerable Sun," on December 25.[273]

Roman legend

It was customary for the Romans to represent divine figures as kings of Latium at the time of their legendary origins. Macrobius states explicitly that the Roman legend of Janus and Saturn is an affabulation, as the true meaning of religious beliefs cannot be openly expressed.[274] In the myth[275] Saturn was the original and autochthonous ruler of the Capitolium, which had thus been called the *Mons Saturnius* in older times and on which once stood the town of *Saturnia*.[276] He was sometimes regarded as the first king of Latium or even the whole of Italy.[277] At the same time, there was a tradition that Saturn had

Figure 78: *Relief held by the Louvre thought to depict the veiled throne of Saturn, either a Roman work of the 1st century AD or a Renaissance copy*

Figure 79: *Alatri's main gate of the cyclopical walls*

been an immigrant god, received by Janus after he was usurped by his son Jupiter and expelled from Greece.[278] In Versnel's view his contradictions—a foreigner with one of Rome's oldest sanctuaries, and a god of liberation who is kept in fetters most of the year—indicate Saturn's capacity for obliterating social distinctions.[279]

The Golden Age of Saturn's reign in Roman mythology differed from the Greek tradition. He arrived in Italy "dethroned and fugitive,"[280] but brought agriculture and civilization for which he was rewarded by Janus with a share of the kingdom, becoming himself king. As the Augustan poet Virgil described it, "He gathered together the unruly race" of fauns and nymphs "scattered over mountain heights, and gave them laws ... . Under his reign were the golden ages men tell of: in such perfect peace he ruled the nations."[281] He was considered the ancestor of the Latin nation as he fathered Picus, the first king of Latium, who married Janus' daughter Canens and in his turn fathered Faunus. Saturn was also said to have founded the five *Saturnian* towns of Latium: Aletrium (today Alatri), Anagnia (Anagni), Arpinum (Arpino), Atina and Ferentinum (Ferentino, also known as Antinum) all located in the Latin Valley, province of Frosinone. All these towns are surrounded by cyclopical walls; their foundation is traditionally ascribed to the Pelasgians.[282]

But Saturn also had a less benevolent aspect, as indicated by the blood shed in his honor during gladiatorial *munera*. His consort in archaic Roman tradition was Lua, sometimes called *Lua Saturni* ("Saturn's Lua") and identified with Lua Mater, "Mother Destruction," a goddess in whose honor the weapons of enemies killed in war were burned, perhaps as expiation.[283] H.S. Versnel, however, proposed that *Lua Saturni* should not be identified with *Lua Mater*, but rather refers to "loosening"; she thus represents the liberating function of Saturn.[284]

Gladiatorial *munera*

Saturn's chthonic nature connected him to the underworld and its ruler Dis Pater, the Roman equivalent of Greek Plouton (Pluto in Latin) who was also a god of hidden wealth.[285] In 3rd-century AD sources and later, Saturn is recorded as receiving gladiatorial offerings *(munera)* during or near the Saturnalia.[286] These gladiator combats, ten days in all throughout December, were presented by the quaestors and sponsored with funds from the treasury of Saturn.[287]

The practice of gladiatorial *munera* was criticized by Christian apologists as a form of human sacrifice.[288] Although there is no evidence of this practice during the Republican era, the offering of gladiators led to later theorizing that the primeval Saturn had demanded human victims. Macrobius says that Dis Pater was placated with human heads and Saturn with sacrificial victims consisting of men *(virorum victimis)*.[289] The figurines that were exchanged as gifts *(sigillaria)* during the Saturnalia may have represented token substitutes.[290]

On coins

In 104 BC, the plebeian tribune Lucius Appuleius Saturninus issued a denarius depicting Saturn driving a four-horse chariot *(quadriga)*, a vehicle associated with rulers, triumphing generals, and sun gods. Saturninus was a popularist

Figure 80: *Saturn driving a quadriga on the reverse of a denarius issued by Saturninus*

politician who had proposed reduced-price grain distribution to the poor of Rome. The head of the goddess Roma appears on the obverse. The Saturnian imagery played on the tribune's name and his intent to alter the social hierarchy to his advantage by basing his political support on the common people *(plebs)* rather than the senatorial elite.[291]

Bibliography

- Georges Dumézil *La religion romaine archaïque* Paris Payot 1974 2nd; Italian translation (expanded version) *La religione romana arcaica* Milano Rizzoli 1977. Edizione e traduzione a cura di Furio Jesi.
- Dominique Briquel "Jupiter, Saturn et le Capitol. Essai de comparaison indoeuropéenne" in *Revue de l' histoire des religions* **198** 2. 1981 p. 131-162.
- Marcel Leglay *Saturn africain. Histoire* **BEFAR** Paris de Boccard 1966.
- H.S. Versnel, "Saturnus and the Saturnalia," in *Inconsistencies in Greek and Roman Religion: Transition and Reversal in Myth and Ritual* (Brill, 1993, 1994), pp. 144–145.

Further reading

> Wikimedia Commons has media related to *Saturnus (deus)*.

- Guirand, Felix (Editor); Aldington, Richard (Translator); Ames, Delano (Translator); & Graves, Robert (Introduction). *New Larousse Encyclopedia of Mythology*. ISBN 0-517-00404-6

External links

> Wikisourcehas the text of the 1911 *Encyclopædia Britannica*article *Saturn (god)*.

- Warburg Institute Iconographic Database (ca 300 images of Saturn)[292]
- (in English) (in Latin) *Flowers of Abu Ma'shar*[293] by Ja'far ibn Muḥammad al-Balkhī depicts and discusses Saturn and his role within astrology, dating from the 9th century.

Modern NASA and ESA probes

Exploration of Saturn

Summary of missions to the outer Solar System

System Spacecraft	Jupiter	Saturn	Uranus	Neptune	Pluto
Pioneer 10	**1973** flyby				
Pioneer 11	**1974** flyby	**1979** flyby			
Voyager 1	**1979** flyby	**1980** flyby			
Voyager 2	**1979** flyby	**1981** flyby	**1986** flyby	**1989** flyby	
Galileo	**1995–2003** orbiter; **1995, 2003** atmospheric				
Ulysses	**1992, 2004** gravity assist				
Cassini–Huygens	**2000** gravity assist	**2004–2017** orbiter; **2005** Titan lander			
New Horizons	**2007** gravity assist				**2015** flyby
Juno	**2016–** orbiter				
Jupiter Icy Moons Explorer	**2022–** Planned orbiter				
Europa Clipper	**2025–** Planned orbiter				

The **exploration of Saturn** has been solely performed by crewless probes. Three missions were flybys, which formed an extended foundation of knowledge about the system. The *Cassini–Huygens* spacecraft, launched in 1997, was in orbit from 2004 to 2017.

Figure 81: *Artwork utilizing exploration data,
as revealed in "Sternstunden" in Oberhausen*

Flybys

Pioneer 11 flyby

Saturn was first visited by *Pioneer 11* in September 1979. It flew within
20,000 km of the top of the planet's cloud layer. Low-resolution images were
acquired of the planet and a few of its moons; the resolution of the images was
not good enough to discern surface features. The spacecraft also studied the
rings; among the discoveries were the thin F-ring and the fact that dark gaps
in the rings are bright when viewed towards the Sun, or in other words, they
are not empty of material. *Pioneer 11* also measured the temperature of Titan
at 250 K.[294]

Voyager

In November 1980, the *Voyager 1* probe visited the Saturn system. It sent
back the first high-resolution images of the planet, rings, and satellites. Surface features of various moons were seen for the first time. Because of the
earlier discovery of a thick atmosphere on Titan, the Voyager controllers at
the Jet Propulsion Laboratory elected for *Voyager 1* to make a close approach
of Titan. This greatly increased knowledge of the atmosphere of the moon, but

Figure 82: *Pioneer 11 image of Saturn.*

also proved that Titan's atmosphere is impenetrable in visible wavelengths, so no surface details were seen. The flyby also changed the spacecraft's trajectory out from the plane of the Solar System which prevented *Voyager 1* from completing the Planetary Grand Tour of Uranus, Neptune and Pluto.

Almost a year later, in August 1981, *Voyager 2* continued the study of the Saturn system. More close-up images of Saturn's moons were acquired, as well as evidence of changes in the rings. *Voyager 2* probed Saturn's upper atmosphere with its radar, to measure temperature and density profiles. *Voyager 2* found that at the highest levels (7 kilopascals pressure) Saturn's temperature was 70 K (–203 °C) (i.e. 70 degrees above absolute zero), while at the deepest levels measured (120 kilopascals) the temperature increased to 143 K (–130 °C). The north pole was found to be 10 K cooler, although this may be seasonal. Unfortunately, during the flyby, the probe's turnable camera platform stuck for a couple of days and some planned imaging was lost. Saturn's gravity was used to direct the spacecraft's trajectory towards Uranus.

The probes discovered and confirmed several new satellites orbiting near or within the planet's rings. They also discovered the small Maxwell and Keeler gaps in the rings.

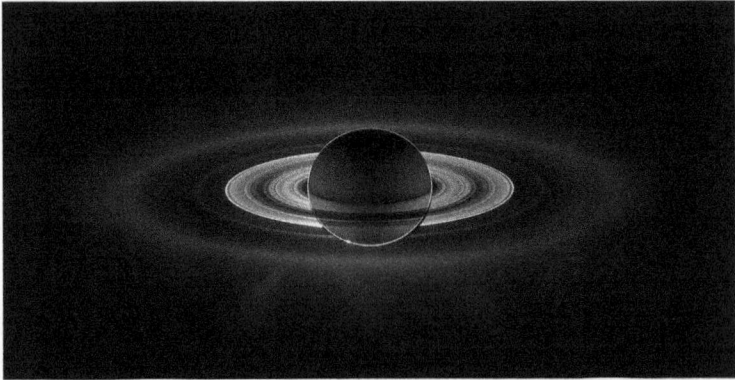

Figure 83: *Saturn eclipses the Sun, as seen from Cassini.*

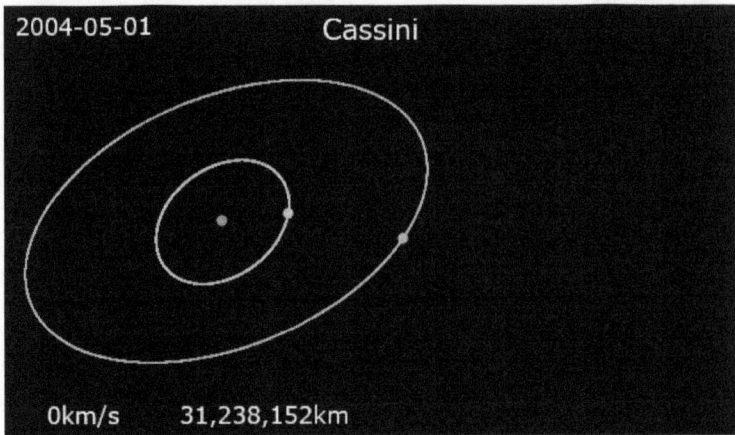

Figure 84: *Animation of Cassini's trajectory around Saturn from 1 May 2004 to 7 August 2008*
Cassini · Saturn · Titan · Iapetus

Cassini orbiter

On July 1, 2004, the *Cassini–Huygens* spacecraft performed the SOI (Saturn Orbit Insertion) maneuver and entered into orbit around Saturn. Before the SOI, *Cassini* had already studied the system extensively. In June 2004, it had conducted a close flyby of Phoebe, sending back high-resolution images and data.

The orbiter completed two Titan flybys before releasing the *Huygens* probe on December 25, 2004. *Huygens* descended onto the surface of Titan on January

Figure 85: *Missions to Saturn face competition from missions to other Solar System bodies*

14, 2005, sending a flood of data during the atmospheric descent and after the landing. During 2005 *Cassini* conducted multiple flybys of Titan and icy satellites.

On March 10, 2006, NASA reported that the *Cassini* probe found evidence of liquid water reservoirs that erupt in geysers on Saturn's moon Enceladus.[295]

On September 20, 2006, a *Cassini* probe photograph revealed a previously undiscovered planetary ring, outside the brighter main rings of Saturn and inside the G and E rings.[296]

In July 2006, *Cassini* saw the first proof of hydrocarbon lakes near Titan's north pole, which was confirmed in January 2007. In March 2007, additional images near Titan's north pole discovered hydrocarbon "seas", the largest of which is almost the size of the Caspian Sea.

In 2009, the probe discovered and confirmed four new satellites. Its primary mission ended in 2008, when the spacecraft completed 74 orbits around the planet. In 2010, the probe began its first extended mission, the *Cassini* Equinox Mission. The *Cassini* Solstice Mission, the second mission extension, lasted through September 2017.[297] The mission ended on September 15, 2017, after a planned atmospheric entry into the planet Saturn.

For the latest information and news releases, see Cassini website[298].

Proposed missions

The Titan Saturn System Mission (**TSSM**) was a joint NASA/ESA proposal for an exploration of Saturn and its moons[299] Titan and Enceladus, where many complex phenomena have been revealed by the recent *Cassini–Huygens* mission. TSSM was competing against the Europa Jupiter System Mission proposal for funding. In February 2009 it was announced that ESA/NASA had given the EJSM mission priority ahead of TSSM,[300,301] although TSSM will continue to be studied for a later launch date. The Titan Saturn System Mission (TSSM) was created by the merging of the ESA's **Titan and Enceladus Mission (TandEM)** with NASA's **Titan Explorer 2007** flagship study.

Other proposed missions to the Saturn system are:

- 2010 JPL: Journey to Enceladus and Titan (JET)
- 2011 Titan Mare Explorer (TiME); an aquatic lander that would explore the methane lakes of the moon Titan. This mission was given US$3 million in May 2011 to develop a detailed concept study as a part of the NASA Discovery program.
- 2012 DLR: Enceladus Explorer (EnEx), a lander with an ice mole.
- 2012 JPL: Life Investigation For Enceladus (LIFE) a sample-return.
- 2015 JPL: Enceladus Life Finder (ELF)[302]

Further reading

- Hannu Karttunen; et al. (1996), *Fundamental Astronomy*[303] (3rd ed.), New York: Springer, ISBN 978-3-540-60936-0, retrieved February 21, 2012.
- NASA/JPL–Caltech (October 18, 2010). "Voyager Science Results: Saturn"[304]. *voyager.jpl.nasa.gov*. Retrieved February 21, 2012.

External links

- NASA's *Cassini* mission to Saturn[305]

Cassini–Huygens spacecraft

Cassini–Huygens

Cassini–Huygens

Artist's concept of *Cassini*'s orbit insertion around Saturn

Mission type	*Cassini*: Saturn orbiter *Huygens*: Titan lander
Operator	*Cassini*: NASA / JPL *Huygens*: ESA / ASI
COSPAR ID	1997-061A[306]
SATCAT no.	25008
Website	• NASA[307] • ESA[308] • ASI[309]

Mission duration	• **Final:**
	  • 19 years, 335 days
	  • 13 years, 76 days at Saturn
	• **En route:**
	  • 6 years, 261 days
	• **Prime mission:**
	  • 3 years
	• **Extended missions:**
	  • Equinox: 2 years, 62 days
	  • Solstice: 6 years, 205 days
	  • Finale: 4 months, 24 days

Spacecraft properties	
Manu-facturer	*Cassini*: Jet Propulsion Laboratory *Huygens*: Alcatel Alenia Space
Launch mass	5,712 kg (12,593 lb)
Dry mass	2,523 kg (5,562 lb)
Power	~885 watts (BOL) ~670 watts (2010) ~663 watts (EOM/2017)

Start of mission	
Launch date	October 15, 1997, 08:43:00 UTC
Rocket	Titan IV(401)B B-33
Launch site	Cape Canaveral SLC-40

End of mission	
Disposal	Controlled entry into Saturn
Last contact	September 15, 2017 • 11:55:39 UTC X-band telemetry • 11:55:46 UTC S-band radio science

Orbital parameters	
Refer-ence system	Kronocentric

Flyby of Venus (Gravity assist)	
Closest approach	April 26, 1998
Distance	283 km (176 mi)

Flyby of Venus (Gravity assist)	
Closest approach	June 24, 1999
Distance	6,052 km (3,761 mi)

Flyby of Earth-Moon system (Gravity assist)	
Closest approach	August 18, 1999, 03:28 UTC
Distance	1,171 km (728 mi)

Flyby of 2685 Masursky (Incidental)	
Closest approach	January 23, 2000

Distance	1,600,000 km (990,000 mi)
Flyby of Jupiter (Gravity assist)	
Closest approach	December 30, 2000
Distance	9,852,924 km (6,122,323 mi)
Saturn orbiter	
Orbital insertion	July 1, 2004, 02:48 UTC
Titan lander	
Spacecraft component	*Huygens*
Landing date	January 14, 2005
Flagship Program	

The *Cassini–Huygens* mission (/kəˈsiːni <wbr />ˈhɔɪɡənz/ *kə-SEE-nee HOY-gənz*), commonly called *Cassini*, was a collaboration between NASA, the European Space Agency (ESA), and the Italian Space Agency (ASI) to send a probe to study the planet Saturn and its system, including its rings and natural satellites. The Flagship-class robotic spacecraft comprised both NASA's *Cassini* probe, and ESA's *Huygens* lander which landed on Saturn's largest moon, Titan. *Cassini* was the fourth space probe to visit Saturn and the first to enter its orbit. The craft were named after astronomers Giovanni Cassini and Christiaan Huygens.

Launched aboard a Titan IVB/Centaur on October 15, 1997, *Cassini* was active in space for nearly 20 years, with 13 years spent orbiting Saturn, studying the planet and its system after entering orbit on July 1, 2004. The voyage to Saturn included flybys of Venus (April 1998 and July 1999), Earth (August 1999), the asteroid 2685 Masursky, and Jupiter (December 2000). Its mission ended on September 15, 2017, when *Cassini's* trajectory took it into Saturn's upper atmosphere and it burned up in order to prevent any risk of contaminating Saturn's moons, which might have offered habitable environments to stowaway terrestrial microbes on the spacecraft. The mission is widely perceived to have been successful beyond expectation. *Cassini-Huygens* has been described by NASA's Planetary Science Division Director as a "mission of firsts",[310] that has revolutionized human understanding of the Saturn system, including its moons and rings, and our understanding of where life might be found in the Solar System.

Cassini's original mission was planned to last for four years, from June 2004 to May 2008. The mission was extended for another two years until September 2010, branded the *Cassini Equinox Mission*. The mission was extended a second and final time with the *Cassini Solstice Mission*, lasting another seven years until September 15, 2017, on which date *Cassini* was de-orbited to burn up in Saturn's upper atmosphere.

The *Huygens* module traveled with *Cassini* until its separation from the probe on December 25, 2004; it was landed by parachute on Titan on January 14, 2005. It returned data to Earth for around 90 minutes, using the orbiter as a relay. This was the first landing ever accomplished in the outer Solar System and the first landing on a moon other than our own.

At the end of its mission, the *Cassini* spacecraft executed the "Grand Finale" of its mission: a number of risky passes through the gaps between Saturn and Saturn's inner rings. The purpose of this phase was to maximize *Cassini*'s scientific outcome before the spacecraft was disposed. The atmospheric entry of *Cassini* ended the mission, but analyses of the returned data will continue for many years.

Overview

Teams from 28 countries made up the joint team responsible for designing, building, flying and collecting data from the *Cassini* orbiter and *Huygens* probe.

The mission was managed by NASA's Jet Propulsion Laboratory in the United States, where the orbiter was assembled. *Huygens* was developed by the European Space Research and Technology Centre. The Centre's prime contractor, Aérospatiale of France (now Thales Alenia Space), assembled the probe with equipment and instruments supplied by many European countries (*Huygens'* batteries and two scientific instruments by the United States). The Italian Space Agency (ASI) provided the *Cassini* orbiter's high-gain radio antenna, with the incorporation of a low-gain antenna (to ensure telecommunications with the Earth for the entire duration of the mission), a compact and lightweight radar, which also uses the high-gain antenna and serves as a synthetic-aperture radar, a radar altimeter, a radiometer, the radio science subsystem (RSS), the visible channel portion VIMS-V of VIMS spectrometer.

The VIMS infrared counterpart was provided by NASA, as well as Main Electronic Assembly, which includes electronic subassemblies provided by CNES of France.

On April 16, 2008, NASA announced a two-year extension of the funding for ground operations of this mission, at which point it was renamed the Cassini Equinox Mission. The round of funding was again extended in February 2010 with the Cassini Solstice Mission.

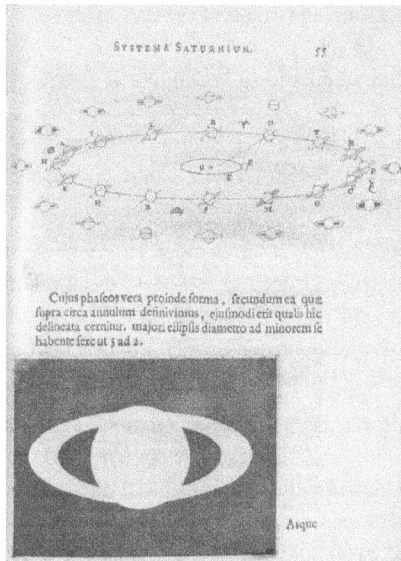

Figure 86: *Huygens' explanation for the aspects of Saturn, Systema Saturnium (1659)*

Naming

The mission consisted of two main elements: the ASI/NASA *Cassini* orbiter, named for the Italian astronomer Giovanni Domenico Cassini, discoverer of Saturn's ring divisions and four of its satellites; and the ESA-developed *Huygens* probe, named for the Dutch astronomer, mathematician and physicist Christiaan Huygens, discoverer of Titan.

The mission was commonly called Saturn Orbiter Titan Probe (SOTP) during gestation, both as a Mariner Mark II mission and generically.

Cassini-Huygens was a Flagship-class mission to the outer planets. The other planetary flagships include *Galileo*, Voyager, and Viking.

Objectives

Cassini had several objectives, including:

• Determining the three-dimensional structure and dynamic behavior of the rings of Saturn.
• Determining the composition of the satellite surfaces and the geological history of each object.

- Determining the nature and origin of the dark material on Iapetus's leading hemisphere.
- Measuring the three-dimensional structure and dynamic behavior of the magnetosphere.
- Studying the dynamic behavior of Saturn's atmosphere at cloud level.
- Studying the time variability of Titan's clouds and hazes.
- Characterizing Titan's surface on a regional scale.

Cassini–Huygens was launched on October 15, 1997, from Cape Canaveral Air Force Station's Space Launch Complex 40 using a U.S. Air Force Titan IVB/Centaur rocket. The complete launcher was made up of a two-stage Titan IV booster rocket, two strap-on solid rocket motors, the Centaur upper stage, and a payload enclosure, or fairing.

The total cost of this scientific exploration mission was about US$3.26 billion, including $1.4 billion for pre-launch development, $704 million for mission operations, $54 million for tracking and $422 million for the launch vehicle. The United States contributed $2.6 billion (80%), the ESA $500 million (15%), and the ASI $160 million (5%). However, these figures are from the press kit which was prepared in October 2000. They do not include inflation over the course of a very long mission, nor do they include the cost of the extended missions.

The primary mission for *Cassini* was completed on July 30, 2008. The mission was extended to June 2010 (*Cassini* Equinox Mission). This studied the Saturn system in detail during the planet's equinox, which happened in August 2009.

On February 3, 2010, NASA announced another extension for *Cassini*, lasting $6^1/_2$ years until 2017, ending at the time of summer solstice in Saturn's northern hemisphere (*Cassini* Solstice Mission). The extension enabled another 155 revolutions around the planet, 54 flybys of Titan and 11 flybys of Enceladus. In 2017, an encounter with Titan changed its orbit in such a way that, at closest approach to Saturn, it was only 3,000 km above the planet's cloudtops, below the inner edge of the D ring. This sequence of "proximal orbits" ended when its final encounter with Titan sent the probe into Saturn's atmosphere to be destroyed.

Itinerary

Selected destinations (ordered by size but not to scale)

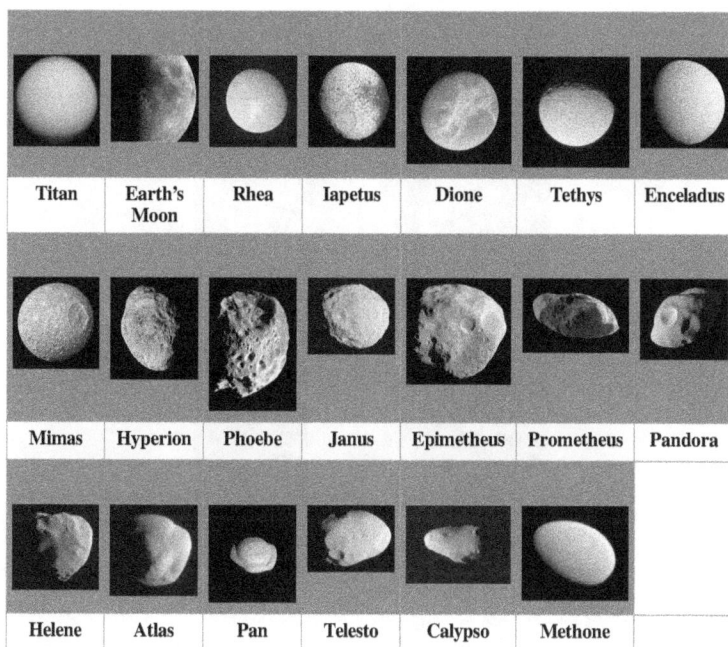

Titan	Earth's Moon	Rhea	Iapetus	Dione	Tethys	Enceladus

Mimas	Hyperion	Phoebe	Janus	Epimetheus	Prometheus	Pandora

Helene	Atlas	Pan	Telesto	Calypso	Methone	

History

Cassini–Huygens's origins date to 1982, when the European Science Foundation and the American National Academy of Sciences formed a working group to investigate future cooperative missions. Two European scientists suggested a paired Saturn Orbiter and Titan Probe as a possible joint mission. In 1983, NASA's Solar System Exploration Committee recommended the same Orbiter and Probe pair as a core NASA project. NASA and the European Space Agency (ESA) performed a joint study of the potential mission from 1984 to 1985. ESA continued with its own study in 1986, while the American astronaut Sally Ride, in her influential 1987 report *NASA Leadership and America's Future in Space*, also examined and approved of the *Cassini* mission.

While Ride's report described the Saturn orbiter and probe as a NASA solo mission, in 1988 the Associate Administrator for Space Science and Applications of NASA, Len Fisk, returned to the idea of a joint NASA and ESA mission. He wrote to his counterpart at ESA, Roger Bonnet, strongly suggesting that ESA choose the *Cassini* mission from the three candidates at hand and promising that NASA would commit to the mission as soon as ESA did.

At the time, NASA was becoming more sensitive to the strain that had developed between the American and European space programs as a result of

Figure 87: *Cassini-Huygens on the launch pad*

European perceptions that NASA had not treated it like an equal during previous collaborations. NASA officials and advisers involved in promoting and planning *Cassini–Huygens* attempted to correct this trend by stressing their desire to evenly share any scientific and technology benefits resulting from the mission. In part, this newfound spirit of cooperation with Europe was driven by a sense of competition with the Soviet Union, which had begun to cooperate more closely with Europe as ESA drew further away from NASA. Late in 1988, ESA chose Cassini–Huygens as its next major mission and the following year the program received major funding in the US.

The collaboration not only improved relations between the two space programs but also helped *Cassini–Huygens* survive congressional budget cuts in the United States. *Cassini–Huygens* came under fire politically in both 1992 and 1994, but NASA successfully persuaded the U.S. Congress that it would be unwise to halt the project after ESA had already poured funds into development because frustration on broken space exploration promises might spill over into other areas of foreign relations. The project proceeded politically smoothly after 1994, although citizens' groups concerned about its potential environmental impact attempted to derail it through protests and lawsuits until and past its 1997 launch.

Figure 88: *Cassini-Huygens assembly*

Spacecraft design

The spacecraft was planned to be the second three-axis stabilized, RTG-powered Mariner Mark II, a class of spacecraft developed for missions beyond the orbit of Mars. *Cassini* was developed simultaneously with the *Comet Rendezvous Asteroid Flyby* (*CRAF*) spacecraft, but budget cuts and project rescopings forced NASA to terminate CRAF development to save *Cassini*. As a result, *Cassini* became more specialized. The Mariner Mark II series was cancelled.

The combined orbiter and probe is the third-largest unmanned interplanetary spacecraft ever successfully launched, behind the Phobos 1 and 2 Mars probes, as well as being among the most complex. The orbiter had a mass of 2,150 kg (4,740 lb), the probe 350 kg (770 lb). With the launch vehicle adapter and 3,132 kg (6,905 lb) of propellants at launch, the spacecraft had a mass of 5,600 kg (12,300 lb).

The *Cassini* spacecraft was 6.8 meters (22 ft) high and 4 meters (13 ft) wide. Spacecraft complexity was increased by its trajectory (flight path) to Saturn, and by the ambitious science at its destination. *Cassini* had 1,630 interconnected electronic components, 22,000 wire connections, and 14 kilometers (8.7 mi) of cabling. The core control computer CPU was a redundant MIL-STD-1750A system. The main propulsion system consisted of one prime and

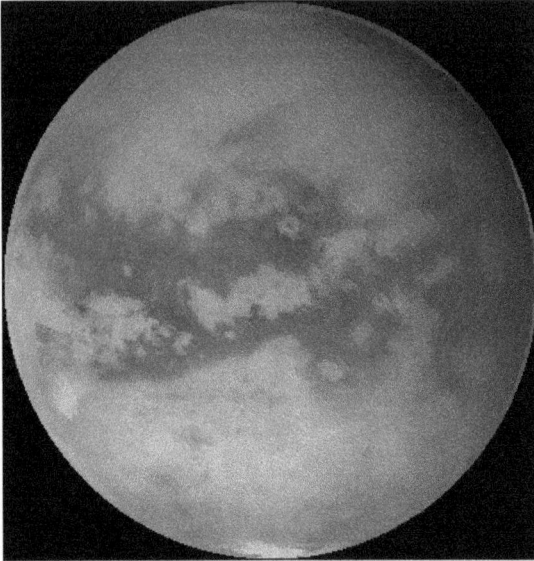

Figure 89: *Titan's surface revealed by VIMS*

one backup R-4D bipropellant rocket engine. The thrust of each engine was 490 newtons and the total spacecraft delta-v was about 2,040 meters per second.[311] Smaller monopropellant rockets provided attitude control.

Cassini was powered by 32.7 kg[312] of plutonium-238—the heat from the material's radioactive decay was turned into electricity. *Huygens* was supported by *Cassini* during cruise, but used chemical batteries when independent.

The probe contained a DVD with more than 616,400 signatures from citizens in 81 countries, collected in a public campaign.

Until September 2017 the *Cassini* probe continued orbiting Saturn at a distance of between 8.2 and 10.2 astronomical units from the Earth. It took 68 to 84 minutes for radio signals to travel from Earth to the spacecraft, and vice versa. Thus ground controllers could not give "real-time" instructions for daily operations or for unexpected events. Even if response were immediate, more than two hours would have passed between the occurrence of a problem and the reception of the engineers' response by the satellite.

Instruments

Summary

Instruments:

Figure 90: *Rhea in front of Saturn*

Figure 91: *Saturn's north polar hexagon*

Figure 92: *Saturn in natural-color (July 2018)*

Figure 93: *Animated 3D model of the spacecraft*

- Optical Remote Sensing ("Located on the remote sensing pallet")
  - Composite Infrared Spectrometer (CIRS)
  - Imaging Science Subsystem (ISS)
  - Ultraviolet Imaging Spectrograph (UVIS)
  - Visible and Infrared Mapping Spectrometer (VIMS)
- Fields, Particles and Waves (mostly in situ)
  - Cassini Plasma Spectrometer (CAPS)
  - Cosmic Dust Analyzer (CDA)
  - Ion and Neutral Mass Spectrometer (INMS)
  - Magnetometer (MAG)
  - Magnetospheric Imaging Instrument (MIMI)
  - Radio and Plasma Wave Science (RPWS)
- Microwave Remote Sensing
  - Radar
  - Radio Science (RSS)

Description

Cassini's instrumentation consisted of: a synthetic aperture radar mapper, a charge-coupled device imaging system, a visible/infrared mapping spectrometer, a composite infrared spectrometer, a cosmic dust analyzer, a radio and plasma wave experiment, a plasma spectrometer, an ultraviolet imaging spectrograph, a magnetospheric imaging instrument, a magnetometer and an ion/neutral mass spectrometer. Telemetry from the communications antenna and other special transmitters (an S-band transmitter and a dual-frequency K_a-band system) was also used to make observations of the atmospheres of Titan and Saturn and to measure the gravity fields of the planet and its satellites.

Cassini Plasma Spectrometer (CAPS)

CAPS was an in situ instrument that measured the flux of charged particles at the location of the spacecraft, as a function of direction and energy. The ion composition was also measured using a time-of-flight mass spectrometer. CAPS measured particles produced by ionisation of molecules originating from Saturn's and Titan's ionosphere, as well as the plumes of Enceladus. CAPS also investigated plasma in these areas, along with the solar wind and its interaction with Saturn's magnetosphere. CAPS was turned off in June 2011, as a precaution due to a "soft" electrical short circuit that occurred in the instrument. It was powered on again in March 2012, but after 78 days another short circuit forced the instrument to be shut down permanently.

Cosmic Dust Analyzer (CDA)

The CDA was an in situ instrument that measured the size, speed, and direction of tiny dust grains near Saturn. It could also measure the grains' chemical elements.[313] Some of these particles orbited Saturn, while others came from other star systems. The CDA on the orbiter was designed to learn more about these particles, the materials in other celestial bodies and potentially about the origins of the universe.

Composite Infrared Spectrometer (CIRS)

The CIRS was a remote sensing instrument that measured the infrared radiation coming from objects to learn about their temperatures, thermal properties, and compositions. Throughout the *Cassini–Huygens* mission, the CIRS measured infrared emissions from atmospheres, rings and surfaces in the vast Saturn system. It mapped the atmosphere of Saturn in three dimensions to determine temperature and pressure profiles with altitude, gas composition, and the distribution of aerosols and clouds. It also measured thermal characteristics and the composition of satellite surfaces and rings.

Ion and Neutral Mass Spectrometer (INMS)

The INMS was an in situ instrument that measured the composition of charged particles (protons and heavier ions) and neutral particles (atoms and molecules) near Titan and Saturn to learn more about their atmospheres. The instrument used a quadrupole mass spectrometer. INMS was also intended to measure the positive ion and neutral environments of Saturn's icy satellites and rings.

Imaging Science Subsystem (ISS)

The ISS was a remote sensing instrument that captured most images in visible light, and also some infrared images and ultraviolet images. The ISS took hundreds of thousands of images of Saturn, its rings, and its moons. The ISS had both a wide-angle camera (WAC) and a narrow-angle camera (NAC). Each of these cameras used a sensitive charge-coupled device (CCD) as its electromagnetic wave detector. Each CCD had a 1,024 square array of pixels, 12 μm on a side. Both cameras allowed for many data collection modes, including on-chip data compression, and were fitted with spectral filters that rotated on a wheel to view different bands within the electromagnetic spectrum ranging from 0.2 to 1.1 μm.

Dual Technique Magnetometer (MAG)

The MAG was an in situ instrument that measured the strength and direction of the magnetic field around Saturn. The magnetic fields are generated partly by the molten core at Saturn's center. Measuring the magnetic field is one of the ways to probe the core. MAG aimed to develop a three-dimensional model of Saturn's magnetosphere, and determine the magnetic state of Titan and its atmosphere, and the icy satellites and their role in the magnetosphere of Saturn.

Magnetospheric Imaging Instrument (MIMI)

The MIMI was both an in situ and remote sensing instrument that produces images and other data about the particles trapped in Saturn's huge magnetic field, or magnetosphere. The in situ component measured energetic ions and electrons while the remote sensing component (the Ion And Neutral Camera, INCA) was an energetic neutral atom imager. This information was used to study the overall configuration and dynamics of the magnetosphere and its interactions with the solar wind, Saturn's atmosphere, Titan, rings, and icy satellites.

Radar

The on-board radar was an active and passive sensing instrument that produced maps of Titan's surface. Radar waves were powerful enough to penetrate the thick veil of haze surrounding Titan. By measuring the send and return time of the signals it is possible to determine the height of large surface features, such as mountains and canyons. The passive radar listened for radio waves that Saturn or its moons may emit.

Radio and Plasma Wave Science instrument (RPWS)

The RPWS was an in situ instrument and remote sensing instrument that receives and measures radio signals coming from Saturn, including the radio waves given off by the interaction of the solar wind with Saturn and Titan. RPWS measured the electric and magnetic wave fields in the interplanetary medium and planetary magnetospheres. It also determined the electron density and temperature near Titan and in some regions of Saturn's magnetosphere using either plasma waves at characteristic frequencies (e.g. the upper hybrid line) or a Langmuir probe. RPWS studied the configuration of Saturn's magnetic field and its relationship to Saturn Kilometric Radiation (SKR), as well as monitoring and mapping Saturn's ionosphere, plasma, and lightning from Saturn's (and possibly Titan's) atmosphere.

Figure 94: *VIMS spectra taken while looking through Titan's atmosphere towards the Sun helped understand the atmospheres of exoplanets (artist's concept; May 27, 2014).*

Radio Science Subsystem (RSS)

The RSS was a remote-sensing instrument that used radio antennas on Earth to observe the way radio signals from the spacecraft changed as they were sent through objects, such as Titan's atmosphere or Saturn's rings, or even behind the Sun. The RSS also studied the compositions, pressures and temperatures of atmospheres and ionospheres, radial structure and particle size distribution within rings, body and system masses and the gravitational field. The instrument used the spacecraft X-band communication link as well as S-band downlink and K_a-band uplink and downlink.

Ultraviolet Imaging Spectrograph (UVIS)

The UVIS was a remote-sensing instrument that captured images of the ultraviolet light reflected off an object, such as the clouds of Saturn and/or its rings, to learn more about their structure and composition. Designed to measure ultraviolet light over wavelengths from 55.8 to 190 nm, this instrument was also a tool to help determine the composition, distribution, aerosol particle content and temperatures of their atmospheres. Unlike other types of spectrometer, this sensitive instrument could take both spectral and spatial readings. It was particularly adept at determining the

Figure 95: *A Cassini RTG before installation*

composition of gases. Spatial observations took a wide-by-narrow view, only one pixel tall and 64 pixels across. The spectral dimension was 1,024 pixels per spatial pixel. It could also take many images that create movies of the ways in which this material is moved around by other forces.

Visible and Infrared Mapping Spectrometer (VIMS)

The VIMS was a remote sensing instrument that captured images using visible and infrared light to learn more about the composition of moon surfaces, the rings, and the atmospheres of Saturn and Titan. It consisted of two cameras - one used to measure visible light, the other infrared. VIMS measured reflected and emitted radiation from atmospheres, rings and surfaces over wavelengths from 350 to 5100 nm, to help determine their compositions, temperatures and structures. It also observed the sunlight and starlight that passes through the rings to learn more about their structure. Scientists used VIMS for long-term studies of cloud movement and morphology in the Saturn system, to determine Saturn's weather patterns.

Plutonium power source

Because of Saturn's distance from the Sun, solar arrays were not feasible as power sources for this space probe. To generate enough power, such arrays

Figure 96: *A glowing-hot plutonium pellet that is the power source of the probe's radioisotope thermoelectric generator*

would have been too large and too heavy. Instead, the *Cassini* orbiter was powered by three radioisotope thermoelectric generators (RTGs), which use heat from the natural decay of about 33 kg (73 lb) of plutonium-238 (in the form of plutonium dioxide) to generate direct current electricity via thermo-electrics. The RTGs on the *Cassini* mission have the same design as those used on the *New Horizons*, *Galileo*, and *Ulysses* space probes, and they were designed to have very long operational lifetimes. At the end of the nominal 11-year *Cassini* mission, they were still able to produce 600 to 700 watts of electrical power. (One of the spare RTGs for the *Cassini* mission was used to power the *New Horizons* mission to Pluto and the Kuiper belt, which was designed and launched later.Wikipedia:Citation needed)

To gain momentum while already in flight, the trajectory of the *Cassini* mission included several gravitational slingshot maneuvers: two fly-by passes of Venus, one more of the Earth, and then one of the planet Jupiter. The terrestrial flyby was the final instance when the probe posed any conceivable danger to human beings. The maneuver was successful, with *Cassini* passing by 1,171 km (728 mi) above the Earth on August 18, 1999. Had there been any malfunction causing the probe to collide with the Earth, NASA's complete environmental impact study estimated that, in the worst case (with an acute angle of entry in which *Cassini* would gradually burn up), a significant fraction of the 33 kg of plutonium-238 inside the RTGs would have been dispersed into the Earth's atmosphere so that up to five billion people (i.e. almost the entire terrestrial population) could have been exposed, causing up to an estimated 5,000 additional cancer deaths over the subsequent decades[314] (0.0005 per cent, i.e. a fraction 0.000005, of a billion cancer deaths expected anyway from other causes; the product is incorrectly calculated elsewhere as 500,000 deaths). However, the chance of this happening were estimated to be less than one in one million.

Telemetry

The *Cassini* spacecraft was capable of transmitting in several different telemetry formats. The telemetry subsystem is perhaps the most important subsystem, because without it there could be no data return.

The telemetry was developed from ground up, due to the spacecraft using a more modern set of computers than previous missions. Therefore, *Cassini* was the first spacecraft to adopt mini-packets to reduce the complexity of the Telemetry Dictionary, and the software development process led to the creation of a Telemetry Manager for the mission.

There were around 1088 channels (in 67 mini-packets) assembled in the *Cassini* Telemetry Dictionary. Out of these 67 lower complexity mini-packets, 6 mini-packets contained the subsystem covariance and Kalman gain elements (161 measurements), not used during normal mission operations. This left 947 measurements in 61 mini-packets.

A total of seven telemetry maps corresponding to 7 AACS telemetry modes were constructed. These modes are: (1) Record; (2) Nominal Cruise; (3) Medium Slow Cruise; (4) Slow Cruise; (5) Orbital Ops; (6) Av; (7) ATE (Attitude Estimator) Calibration. These 7 maps cover all spacecraft telemetry modes.

Huygens probe

<templatestyles src="Multiple_image/styles.css" />

Huygens view of Titan's surface

Same image with different data processing

The *Huygens* probe, supplied by the European Space Agency (ESA) and named after the 17th century Dutch astronomer who first discovered Titan, Christiaan Huygens, scrutinized the clouds, atmosphere, and surface of Saturn's moon Titan in its descent on January 15, 2005. It was designed to enter and brake in Titan's atmosphere and parachute a fully instrumented robotic laboratory down to the surface.[315]

The probe system consisted of the probe itself which descended to Titan, and the probe support equipment (PSE) which remained attached to the orbiting spacecraft. The PSE includes electronics that track the probe, recover the data gathered during its descent, and process and deliver the data to the orbiter that transmits it to Earth. The core control computer CPU was a redundant MIL-STD-1750A control system.

The data were transmitted by a radio link between *Huygens* and *Cassini* provided by Probe Data Relay Subsystem (PDRS). As the probe's mission could not be telecommanded from Earth because of the great distance, it was automatically managed by the Command Data Management Subsystem (CDMS). The PDRS and CDMS were provided by the Italian Space Agency (ASI).

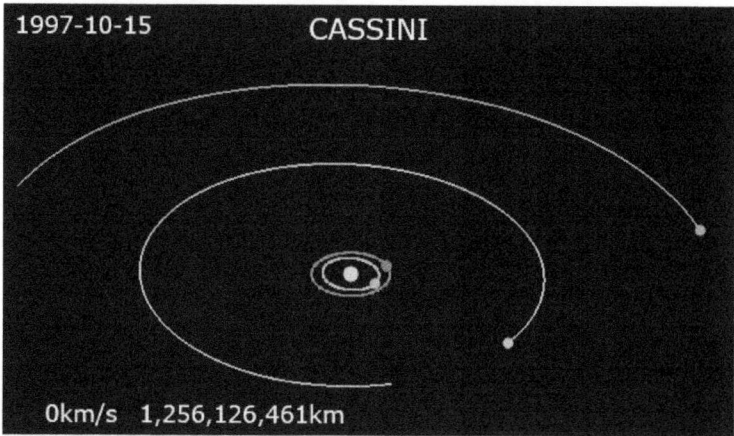

Figure 97: *Animation of Cassini's trajectory from 15 October 1997 to 4 May 2008*
Cassini–Huygens · Jupiter · Saturn · Earth · Venus

After *Cassini*'s launch, it was discovered that data sent back from the probe to European Space Agency's mission control center was largely unreadable. It was determined that *Cassini*'s receiver was unable to accurately process the changes in frequency and wavelength of the signal it would be receiving from *Huygens* during its descent to Titan. The problem was rectified by changing the distance and angle of *Cassini*'s trajectory to *Huygens* during the landing.[316]

Selected events and discoveries

Venus and Earth fly-bys and the cruise to Jupiter

The *Cassini* space probe performed two gravitational-assist flybys of Venus on April 26, 1998, and June 24, 1999. These flybys provided the space probe with enough momentum to travel all the way out to the asteroid belt. At that point, the Sun's gravity pulled the space probe back into the inner Solar System.

On August 18, 1999, at 03:28 UTC, the craft made a gravitational-assist flyby of the Earth. One hour and 20 minutes before closest approach, *Cassini* made its closest approach to the Earth's Moon at 377,000 kilometers, and it took a series of calibration photos.

On January 23, 2000, *Cassini* performed a flyby of the asteroid 2685 Masursky at around 10:00 UTC. It took photos in the period five to seven hours before the flyby at a distance of 1.6 million kilometers, and a diameter of 15 to 20 km was estimated for the asteroid.

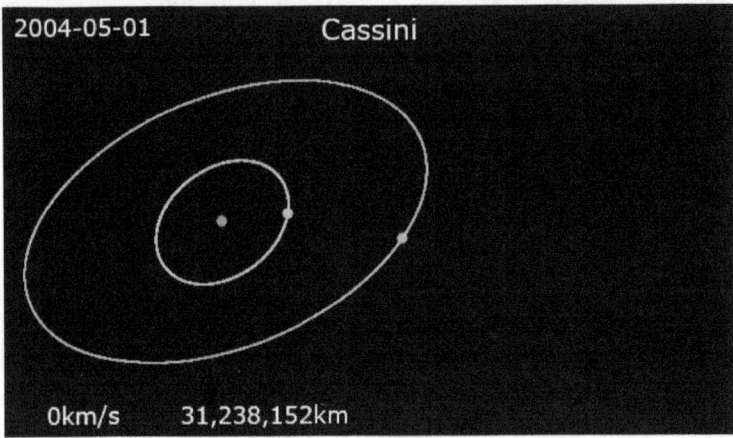

Figure 98: *Animation of Cassini's trajectory around Saturn from 1 May 2004 to 7 August 2008*
Cassini · Saturn · Titan · Iapetus

Figure 99: *Picture of the Moon during flyby*

Figure 100: *A Jupiter flyby picture*

Jupiter flyby

Cassini made its closest approach to Jupiter on December 30, 2000, and made many scientific measurements. About 26,000 images of Jupiter, its faint rings, and its moons were taken during the six month flyby. It produced the most detailed global color portrait of the planet yet (see image at right), in which the smallest visible features are approximately 60 km (37 mi) across.

A major finding of the flyby, announced on March 6, 2003, was of Jupiter's atmospheric circulation. Dark "belts" alternate with light "zones" in the atmosphere, and scientists had long considered the zones, with their pale clouds, to be areas of upwelling air, partly because many clouds on Earth form where air is rising. But analysis of *Cassini* imagery showed that individual storm cells of upwelling bright-white clouds, too small to see from Earth, pop up almost without exception in the dark belts. According to Anthony Del Genio of NASA's Goddard Institute for Space Studies, "the belts must be the areas of net-rising atmospheric motion on Jupiter, [so] the net motion in the zones has to be sinking."

Other atmospheric observations included a swirling dark oval of high atmospheric-haze, about the size of the Great Red Spot, near Jupiter's north pole. Infrared imagery revealed aspects of circulation near the poles, with

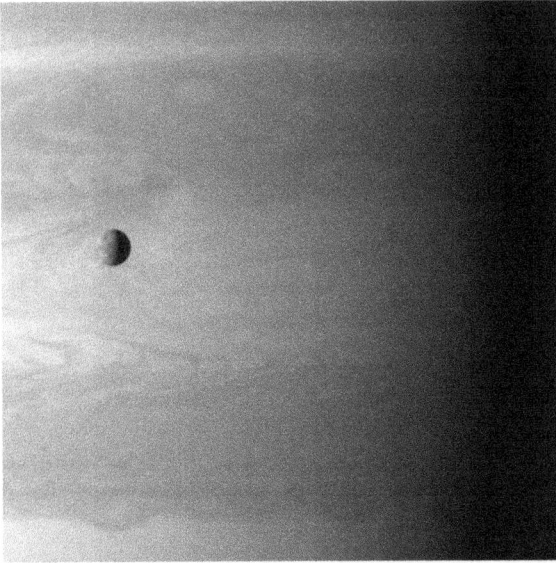

Figure 101: *Cassini photographed Io transiting Jupiter on January 1, 2001.*

bands of globe-encircling winds, with adjacent bands moving in opposite directions.

The same announcement also discussed the nature of Jupiter's rings. Light scattering by particles in the rings showed the particles were irregularly shaped (rather than spherical) and likely originate as ejecta from micrometeorite impacts on Jupiter's moons, probably Metis and Adrastea.

Tests of general relativity

On October 10, 2003, the mission's science team announced the results of tests of Albert Einstein's general theory of relativity, performed by using radio waves transmitted from the *Cassini* space probe. The radio scientists measured a frequency shift in the radio waves to and from the spacecraft, as those passed close to the Sun. According to the general theory of relativity, a massive object like the Sun causes space-time to curve, causing a beam of radiowaves (or light, or any form of electromagnetic radiation) that passes by the Sun to travel farther (known as the Shapiro time delay).

Although some measurable deviations from the values calculated using the general theory of relativity are predicted by some unusual cosmological models, no such deviations were found by this experiment. Previous tests using radiowaves transmitted by the *Viking* and *Voyager* space probes were in

Figure 102: *The possible formation of a new moon was captured on April 15, 2013.*

agreement with the calculated values from general relativity to within an accuracy of one part in one thousand. The more refined measurements from the *Cassini* space probe experiment improved this accuracy to about one part in 51,000.[317] The data firmly support Einstein's general theory of relativity.Wikipedia:Citation needed

New moons of Saturn

In total, the *Cassini* mission discovered seven new moons orbiting Saturn.[318] Using images taken by *Cassini*, researchers discovered Methone, Pallene and Polydeuces in 2004, although later analysis revealed that Voyager 2 had photographed Pallene in its 1981 flyby of the ringed planet.

On May 1, 2005, a new moon was discovered by *Cassini* in the Keeler gap. It was given the designation S/2005 S 1 before being named Daphnis. A fifth new moon was discovered by *Cassini* on May 30, 2007, and was provisionally labeled S/2007 S 4. It is now known as Anthe. A press release on February 3, 2009 showed a sixth new moon found by *Cassini*. The moon is approximately 1/3 of a mile (500 m) in diameter within the G-ring of the ring system of Saturn, and is now named Aegaeon (formerly S/2008 S 1). A press release on November 2, 2009 mentions the seventh new moon found by *Cassini* on

Figure 103: *Discovery photograph of moon Daphnis*

July 26, 2009. It is presently labeled S/2009 S 1 and is approximately 300 m (1000 ft) in diameter in the B-ring system.

On April 14, 2014, NASA scientists reported the possible beginning of a new moon in Saturn's A Ring.

Phoebe flyby

On June 11, 2004, *Cassini* flew by the moon Phoebe. This was the first opportunity for close-up studies of this moon (Voyager 2 performed a distant flyby in 1981 but returned no detailed images). It also was *Cassini's* only possible flyby for Phoebe due to the mechanics of the available orbits around Saturn.

The first close-up images were received on June 12, 2004, and mission scientists immediately realized that the surface of Phoebe looks different from asteroids visited by spacecraft. Parts of the heavily cratered surface look very bright in those pictures, and it is currently believed that a large amount of water ice exists under its immediate surface.

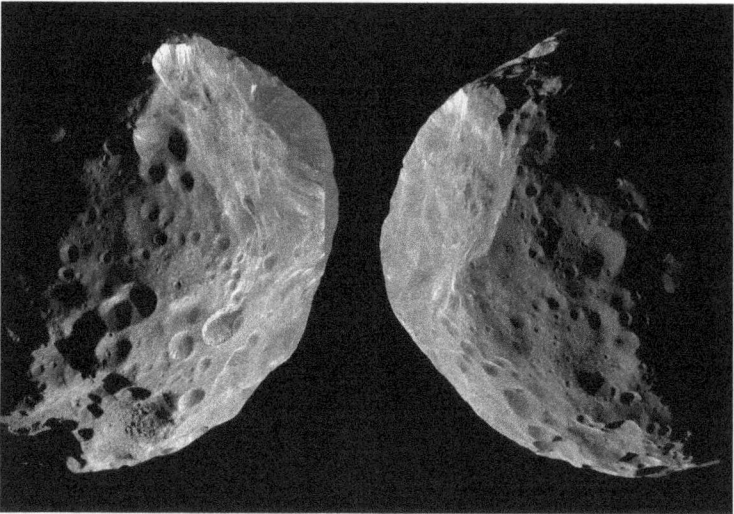

Figure 104: *Cassini arrival (left) and departure mosaics of Phoebe (2004)*

Saturn rotation

In an announcement on June 28, 2004, *Cassini* program scientists described the measurement of the rotational period of Saturn. Because there are no fixed features on the surface that can be used to obtain this period, the repetition of radio emissions was used. This new data agreed with the latest values measured from Earth, and constituted a puzzle to the scientists. It turns out that the radio rotational period had changed since it was first measured in 1980 by *Voyager 1*, and it was now 6 minutes longer. This, however, does not indicate a change in the overall spin of the planet. It is thought to be due to variations in the upper atmosphere and ionosphere at the latitudes which are magnetically connected to the radio source region.

Orbiting Saturn

On July 1, 2004, the spacecraft flew through the gap between the F and G rings and achieved orbit, after a seven-year voyage. It was the first spacecraft to ever orbit Saturn.

The Saturn Orbital Insertion (SOI) maneuver performed by *Cassini* was complex, requiring the craft to orient its High-Gain Antenna away from Earth and along its flight path, to shield its instruments from particles in Saturn's rings. Once the craft crossed the ring plane, it had to rotate again to point its engine along its flight path, and then the engine fired to decelerate the craft by

Figure 105: *Saturn reached equinox in 2008,*
shortly after the end of the prime mission.

622 meters/s to allow Saturn to capture it. *Cassini* was captured by Saturn's gravity at around 8:54 pm Pacific Daylight Time on June 30, 2004. During the maneuver *Cassini* passed within 20,000 km (12,000 mi) of Saturn's cloud tops.

When Cassini was in Saturnian orbit, departure from the Saturn system was evaluated in 2008 during end of mission planning.Wikipedia:Please clarify

Titan flybys

Cassini had its first flyby of Saturn's largest moon, Titan, on July 2, 2004, a day after orbit insertion, when it approached to within 339,000 km (211,000 mi) of Titan. Images taken through special filters (able to see through the moon's global haze) showed south polar clouds thought to be composed of methane and surface features with widely differing brightness. On October 27, 2004, the spacecraft executed the first of the 45 planned close flybys of Titan when it passed a mere 1,200 kilometers above the moon. Almost four gigabits of data were collected and transmitted to Earth, including the first radar images of the moon's haze-enshrouded surface. It revealed the surface of Titan (at least the area covered by radar) to be relatively level, with topography reaching no more than about 50 meters in altitude. The flyby provided a remarkable increase in imaging resolution over previous coverage. Images with up to 100 times better resolution were taken and are typical of resolutions planned for subsequent Titan flybys. Cassini collected pictures of Titan and the lakes of methane were similar to the lakes of Earth.

Figure 106:
Titan – infrared views (2004 – 2017)

Huygens lands on Titan

🔍 Raw images from the *Huygens* probe descent on 14 January 2005 (37 pages)[319]

© ESA/NASA/JPL/U. of Arizona. (ESA hosting)

Cassini released the *Huygens* probe on December 25, 2004, by means of a spring and spiral rails intended to rotate the probe for greater stability. It entered the atmosphere of Titan on January 14, 2005, and after a two-and-a-half-hour descent landed on solid ground. Although *Cassini* successfully relayed 350 of the pictures that it received from *Huygens* of its descent and landing site, a software error failed to turn on one of the *Cassini* receivers and caused the loss of another 350 pictures. While landing, for caution, NASA loaded Huygens with 3 parachutes.

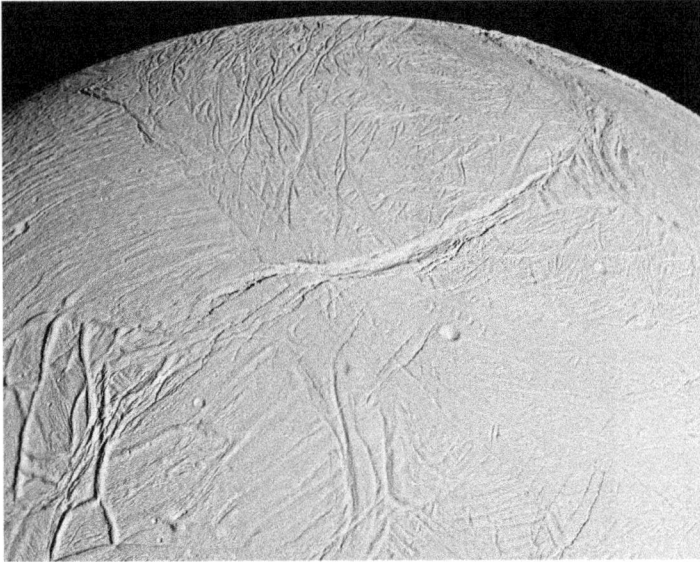

Figure 107: *View of Enceladus's Europa-like surface with the Labtayt Sulci fractures at center and the Ebony (left) and Cufa dorsa at lower left; imaged by Cassini on February 17, 2005*

Enceladus flybys

During the first two close flybys of the moon Enceladus in 2005, *Cassini* discovered a deflection in the local magnetic field that is characteristic for the existence of a thin but significant atmosphere. Other measurements obtained at that time point to ionized water vapor as its main constituent. *Cassini* also observed water ice geysers erupting from the south pole of Enceladus, which gives more credibility to the idea that Enceladus is supplying the particles of Saturn's E ring. Mission scientists began to suspect that there may be pockets of liquid water near the surface of the moon that fuel the eruptions.

On March 12, 2008, *Cassini* made a close fly-by of Enceladus, passing within 50 km of the moon's surface.[320] The spacecraft passed through the plumes extending from its southern geysers, detecting water, carbon dioxide and various hydrocarbons with its mass spectrometer, while also mapping surface features that are at much higher temperature than their surroundings with the infrared spectrometer.[321] *Cassini* was unable to collect data with its cosmic dust analyzer due to an unknown software malfunction.

Wikinews has related news: *Cassini discovers organic material on Saturn moon*

On November 21, 2009, *Cassini* made its eighth flyby of Enceladus, this time with a different geometry, approaching within 1,600 kilometers (990 mi) of the surface. The Composite Infrared Spectrograph (CIRS) instrument produced a map of thermal emissions from the Baghdad Sulcus 'tiger stripe'. The data returned helped create a detailed and high resolution mosaic image of the southern part of the moon's Saturn-facing hemisphere.

On April 3, 2014, nearly ten years after *Cassini* entered Saturn's orbit, NASA reported evidence of a large salty internal ocean of liquid water in Enceladus. The presence of an internal salty ocean in contact with the moon's rocky core, places Enceladus "among the most likely places in the Solar System to host alien microbial life". On June 30, 2014, NASA celebrated ten years of *Cassini* exploring Saturn and its moons, highlighting the discovery of water activity on Enceladus among other findings.

In September 2015, NASA announced that gravitational and imaging data from *Cassini* were used to analyze the librations of Enceladus' orbit and determined that the moon's surface is not rigidly joined to its core, concluding that the underground ocean must therefore be global in extent.

On October 28, 2015, *Cassini* performed a close flyby of Enceladus, coming within 49 km (30 mi) of the surface, and passing through the icy plume above the south pole.

Radio occultations of Saturn's rings

In May 2005, *Cassini* began a series of radio occultation experiments, to measure the size-distribution of particles in Saturn's rings, and measure the atmosphere of Saturn itself. For over four months, the craft completed orbits designed for this purpose. During these experiments, it flew behind the ring plane of Saturn, as seen from Earth, and transmitted radio waves through the particles. The radio signals received on Earth were analyzed, for frequency, phase, and power shift of the signal to determine the structure of the rings.

File:Saturn's rings in visible light and radio.jpg

Upper image: visible color mosaic of Saturn's rings taken on December 12, 2004. Lower image: simulated view constructed from a radio occultation observation on May 3, 2005. Color in the lower image represents ring particle sizes.

Figure 108: *Ligeia Mare, on the left, is compared at scale to Lake Superior.*

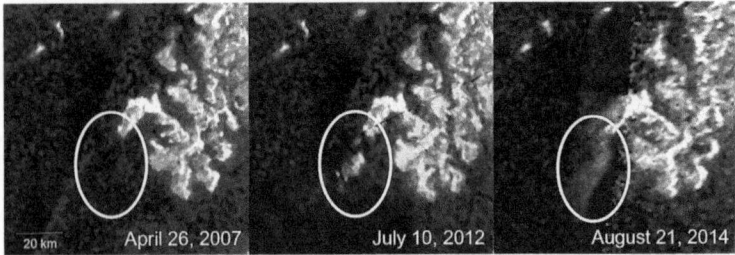

Figure 109: *Titan - Evolving feature in Ligeia Mare (August 21, 2014).*

Spokes in rings verified

In images captured September 5, 2005, *Cassini* detected spokes in Saturn's rings, previously seen only by the visual observer Stephen James O'Meara in 1977 and then confirmed by the Voyager space probes in the early 1980s.Wikipedia:Citing sources#What information to include

Lakes of Titan

Radar images obtained on July 21, 2006 appear to show lakes of liquid hydrocarbon (such as methane and ethane) in Titan's northern latitudes. This is the first discovery of currently existing lakes anywhere besides on Earth. The lakes range in size from one to one-hundred kilometers across.

On March 13, 2007, the Jet Propulsion Laboratory announced that it had found strong evidence of seas of methane and ethane in the northern hemisphere of

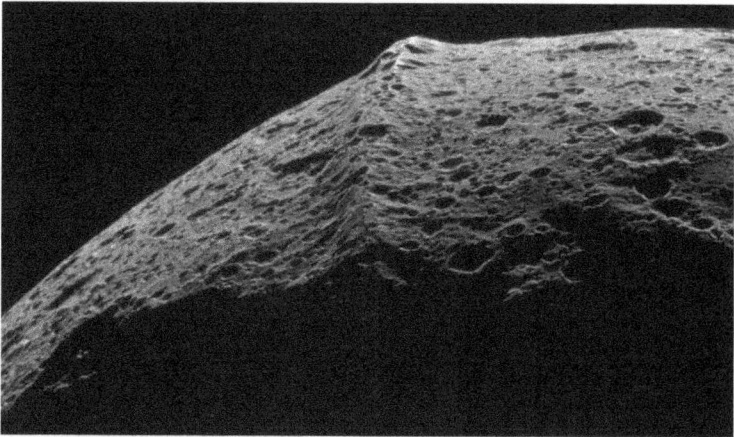

Figure 110: *Taken on September 10, 2007 at a distance of 62,331 km (38,731 mi) Iapetus's equatorial ridge and surface are revealed. (CL1 and CL2 filters)*

Titan. At least one of these is larger than any of the Great Lakes in North America.

Saturn hurricane

In November 2006, scientists discovered a storm at the south pole of Saturn with a distinct eyewall. This is characteristic of a hurricane on Earth and had never been seen on another planet before. Unlike a terrestrial hurricane, the storm appears to be stationary at the pole. The storm is 8,000 kilometers (5,000 mi) across, and 70 kilometers (43 mi) high, with winds blowing at 560 kilometers per hour (350 mph).

Iapetus flyby

On September 10, 2007, *Cassini* completed its flyby of the strange, two-toned, walnut-shaped moon, Iapetus. Images were taken from 1,000 miles (1,600 km) above the surface. As it was sending the images back to Earth, it was hit by a cosmic ray that forced it to temporarily enter safe mode. All of the data from the flyby were recovered.

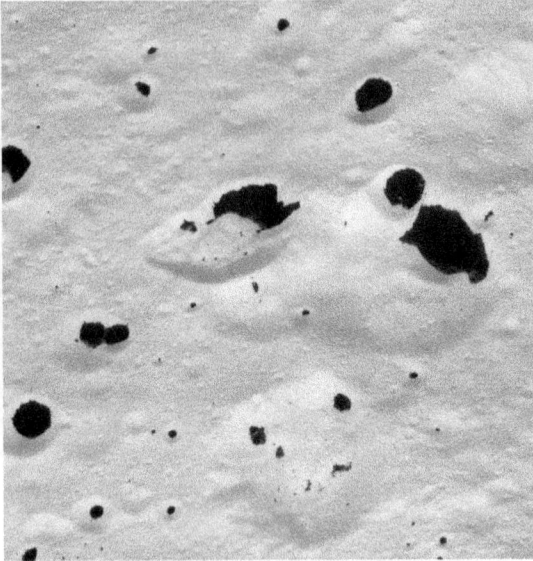

Figure 111: *Closeup of Iapetus surface, 2007*

Mission extension

On April 15, 2008, *Cassini* received funding for a 27-month extended mission. It consisted of 60 more orbits of Saturn, with 21 more close Titan flybys, seven of Enceladus, six of Mimas, eight of Tethys, and one targeted flyby each of Dione, Rhea, and Helene. The extended mission began on July 1, 2008, and was renamed the **Cassini Equinox Mission** as the mission coincided with Saturn's equinox.

Second mission extension

A proposal was submitted to NASA for a second mission extension (September 2010 – May 2017), provisionally named the extended-extended mission or XXM. This ($60M pa) was approved in February 2010 and renamed the **Cassini Solstice Mission**.[322] It included *Cassini* orbiting Saturn 155 more times, conducting 54 additional flybys of Titan and 11 more of Enceladus.

Figure 112: *Northern hemisphere storm in 2011*

Great Storm of 2010 and aftermath

On October 25, 2012, *Cassini* witnessed the aftermath of the massive Great White Spot storm that recurs roughly every 30 years on Saturn. Data from the composite infrared spectrometer (CIRS) instrument indicated a powerful discharge from the storm that caused a temperature spike in the stratosphere of Saturn 83 K (83 °C; 149 °F) above normal. Simultaneously, a huge increase in ethylene gas was detected by NASA researchers at Goddard Research Center in Greenbelt, Maryland. Ethylene is a colorless gas that is highly uncommon on Saturn and is produced both naturally and through man-made sources on Earth. The storm that produced this discharge was first observed by the spacecraft on December 5, 2010 in Saturn's northern hemisphere. The storm is the first of its kind to be observed by a spacecraft in orbit around Saturn as well as the first to be observed at thermal infrared wavelengths, allowing scientists to observe the temperature of Saturn's atmosphere and track phenomena that are invisible to the naked eye. The spike of ethylene gas that was produced by the storm reached levels that were 100 times more than those thought possible for Saturn. Scientists have also determined that the storm witnessed was the largest, hottest stratospheric vortex ever detected in the Solar System, initially being larger than Jupiter's Great Red Spot.

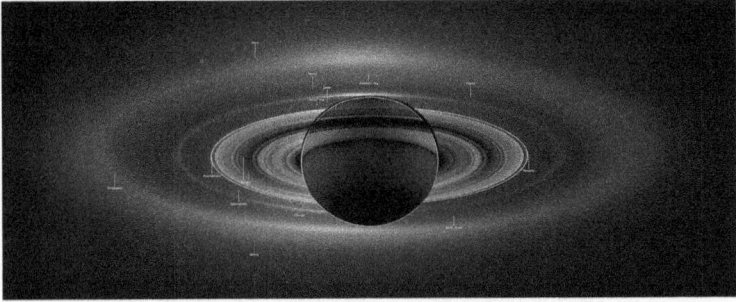

Figure 113: *The Day the Earth Smiled - Saturn with some of its moons, Earth, Venus, and Mars as visible in this Cassini montage (July 19, 2013)*

Venus transit

On December 21, 2012, *Cassini* observed a transit of Venus across the Sun. The VIMS instrument analyzed sunlight passing through the Venusian atmosphere. VIMS previously observed the transit of exoplanet HD 189733 b.

The Day the Earth Smiled

On July 19, 2013, the probe was pointed towards Earth to capture an image of the Earth and the Moon, as part of a natural light, multi-image portrait of the entire Saturn system. The event was unique as it was the first time NASA informed the public that a long-distance photo was being taken in advance. The imaging team said they wanted people to smile and wave to the skies, with *Cassini* scientist Carolyn Porco describing the moment as a chance to "celebrate life on the Pale Blue Dot".

Rhea flyby

On February 10, 2015, the *Cassini* spacecraft visited Rhea more closely, coming within 47,000 km (29,000 mi). The spacecraft observed the moon with its cameras producing some of the highest resolution color images yet of Rhea.

Hyperion flyby

Cassini performed its latest flyby of Saturn's moon Hyperion on May 31, 2015, at a distance of about 34,000 km (21,000 mi).

<templatestyles src="Multiple_image/styles.css" />

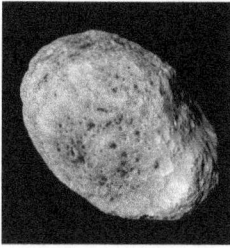

Hyperion - context view

from 37,000 km (23,000 mi)

(May 31, 2015)

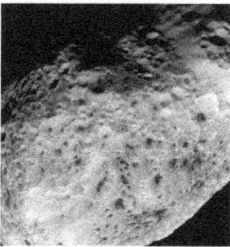

Hyperion - close-up view

from 38,000 km (24,000 mi)

(May 31, 2015)

Dione flyby

Cassini performed its last flyby of Saturn's moon Dione on August 17, 2015, at a distance of about 475 km (295 mi). A previous flyby was performed on June 16.

Hexagon changes color

Between 2012 and 2016, the persistent hexagonal cloud pattern at Saturn's north pole changed from a mostly blue color to more of a golden color. One theory for this is a seasonal change: extended exposure to sunlight may be creating haze as the pole swivels toward the sun. It was previously noted that there was less blue color overall on Saturn between 2004 and 2008.

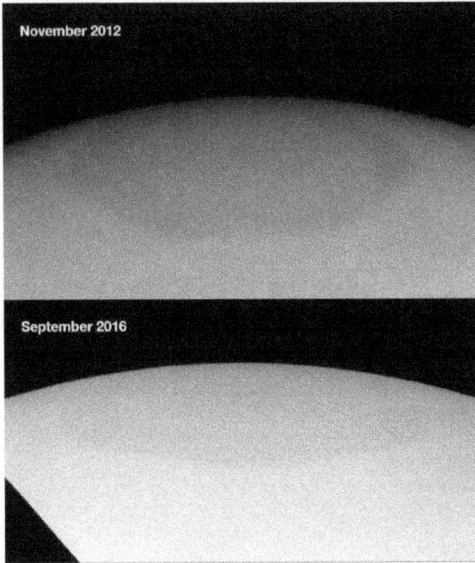

Figure 114:
2012 and 2016:
hexagon color changes

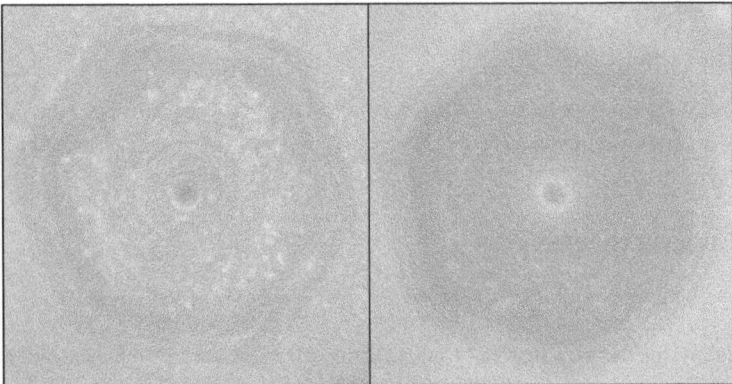

Figure 115:
2013 and 2017:
hexagon color changes

Grand Finale and destruction

Cassini's end involved a series of close Saturn passes, approaching within the rings, then an entry into Saturn's atmosphere on September 15, 2017, to destroy the spacecraft. This method was chosen because it is imperative to ensure protection and prevent biological contamination to any of the moons of Saturn thought to offer potential habitability.

In 2008 a number of options were evaluated to achieve this goal, each with varying funding, scientific, and technical challenges. A short period Saturn impact for an end of mission was rated "excellent" with the reasons "D-ring option satisfies unachieved AO goals; cheap and easily achievable" while collision with an icy moon was rated "good" for being "cheap and achievable anywhere/time".

On November 29, 2016, the spacecraft performed a Titan flyby that took it to the gateway of F-ring orbits: This was the start of the Grand Finale phase culminating in its impact with the planet. A final Titan flyby on April 22, 2017, changed the orbit again to fly through the gap between Saturn and its inner ring days later on April 26. *Cassini* passed about 3,100 km (1,900 mi) above Saturn's cloud layer and 320 km (200 mi) from the visible edge of the inner ring; it successfully took images of Saturn's atmosphere and began returning data the next day. After a further 22 orbits through the gap, the mission was ended with a dive into Saturn's atmosphere on September 15; signal was lost at 7:55:46 AM EDT on September 15, 2017, just 30 seconds later than predicted. It is estimated that the spacecraft burned up about 45 seconds after the last transmission.

<templatestyles src="Multiple_image/styles.css" />

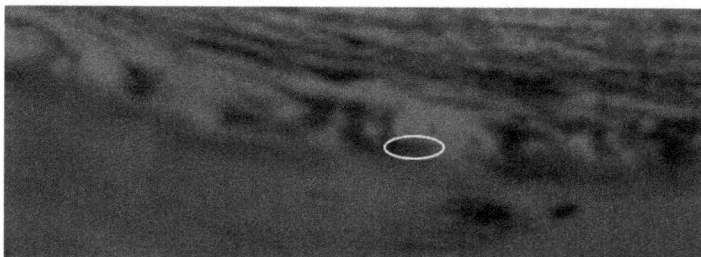

Cassini impact site on Saturn (visual/IR mapping spectrometer; September 15, 2017)

<templatestyles src="Multiple_image/styles.css" />

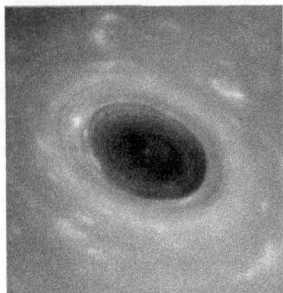

A close-up image of Saturn's atmosphere from about 3,100 km (1,900 mi) above the cloud layer, taken by *Cassini* on its first dive on April 26, 2017, at the start of the Grand Finale

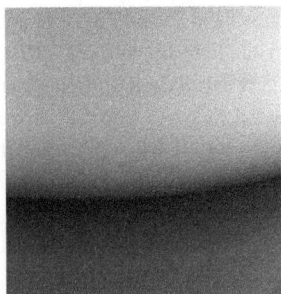

Last image (color) taken by *Cassini* as it descended toward Saturn. The image was taken 634,000 km (394,000 mi) above Saturn on September 14, 2017, at 19:59 UTC.

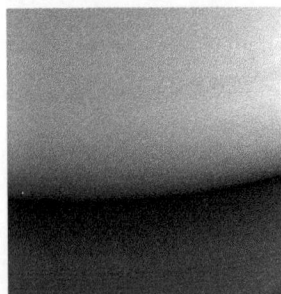

Last image (b&w) taken by the imaging cameras on the *Cassini* spacecraft (September 14, 2017, at 19:59 UTC)

<templatestyles src="Multiple_image/styles.css" />

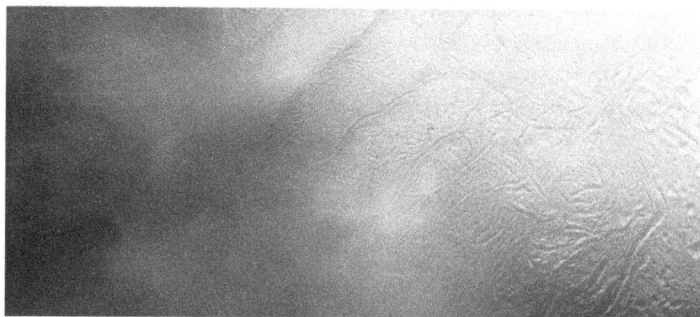

Video (03:40) detailing *Cassini*'s Grand Finale mission
and a look back at what the mission has accomplished.

Missions

The spacecraft operation was organized around a series of missions. Each is structured according to a certain amount of funding, goals, etc. At least 260 scientists from 17 countries have worked on the *Cassini–Huygens* mission; in addition thousands of people overall worked to design, manufacture, and launch the mission.

- Prime Mission, July 2004 through June 2008.
- *Cassini* Equinox Mission was a two-year mission extension which ran from July 2008 through September 2010.
- *Cassini* Solstice Mission ran from October 2010 through April 2017. (Also known as the XXM mission.)
- Grand Finale (spacecraft directed into Saturn), April 2017 to September 15, 2017.

Glossary

- AACS: Attitude and Articulation Control Subsystem
- ACS: Attitude Control Subsystem
- AFC: AACS Flight Computer
- ARWM: Articulated Reaction Wheel Mechanism
- ASI: Agenzia Spaziale Italiana, the Italian space agency
- BIU: Bus Interface Unit
- BOL: Beginning of Life
- CAM: Command Approval Meeting
- CDS: Command and Data Subsystem—Cassini computer that commands and collects data from the instruments

- CICLOPS: Cassini Imaging Central Laboratory for Operations[323]
- CIMS: *Cassini* Information Management System
- CIRS: Composite Infrared Spectrometer
- DCSS: Descent Control Subsystem
- DSCC: Deep Space Communications Center
- DSN: Deep Space Network (large antennas around the Earth)
- DTSTART: Dead Time Start
- ELS: Electron Spectrometer (part of CAPS instrument)
- EOM: End of Mission
- ERT: Earth-received time, UTC of an event
- ESA: European Space Agency
- ESOC: European Space Operations Centre
- FSW: flight software
- HGA: High Gain Antenna
- HMCS: *Huygens* Monitoring and Control System
- HPOC: *Huygens* Probe Operations Center
- IBS: Ion Beam Spectrometer (part of CAPS instrument)
- IEB: Instrument Expanded Blocks (instrument command sequences)
- IMS: Ion Mass Spectrometer (part of CAPS instrument)
- ITL: Integrated Test Laboratory—spacecraft simulator
- IVP: Inertial Vector Propagator
- LGA: Low Gain Antenna
- NAC: Narrow Angle Camera
- NASA: National Aeronautics and Space Administration, the United States of America space agency
- OTM: Orbit Trim Maneuver
- PDRS: Probe Data Relay Subsystem
- PHSS: Probe Harness SubSystem
- POSW: Probe On-Board Software
- PPS: Power and Pyrotechnic Subsystem
- PRA: Probe Relay Antenna
- PSA: Probe Support Avionics
- PSIV: Preliminary Sequence Integration and Validation
- PSE: probe support equipment
- RCS: Reaction Control System
- RFS: Radio Frequency Subsystem
- RPX: ring plane crossing
- RWA: Reaction Wheel Assembly
- SCET: Spacecraft Event Time
- SCR: sequence change requests
- SKR: Saturn Kilometric Radiation
- SOI: Saturn Orbit Insertion (July 1, 2004)

- SOP: Science Operations Plan
- SSPS: Solid State Power Switch
- SSR: Solid State Recorder
- SSUP: Science and Sequence Update Process
- TLA: Thermal Louver Assemblies
- USO: UltraStable Oscillator
- VRHU: Variable Radioisotope Heater Units
- WAC: Wide Angle Camera
- XXM: Extended-Extended Mission

Further reading

- Ralph Lorenz (2017). *NASA/ESA/ASI Cassini-Huygens: 1997 onwards (Cassini orbiter, Huygens probe and future exploration concepts) (Owners' Workshop Manual)*. Haynes Manuals, UK. ISBN 978-1785211119.
- Karl Grossman (1997). *The Wrong Stuff: The Space Program's Nuclear Threat to Our Planet*. Common Courage Press. ISBN 1-56751-125-2.
- David M. Harland (2002). *Mission to Saturn: Cassini and the Huygens Probe*. Springer-Verlag. ISBN 1-85233-656-0.
- Ralph Lorenz; Jacqueline Mitton (2002). *Lifting Titan's Veil: Exploring the Giant Moon of Saturn*. Cambridge University Press. ISBN 0-521-79348-3.
- Meltzer, Michael (2015). *The Cassini-Huygens Visit to Saturn: A Historic Mission to the Ringed Planet*. Cham: Springer International Publishing Switzerland. ISBN 978-3-319-07608-9.
- Irene Klotz (August 31, 2017). "Cassini's Ringside Seat At Saturn Coming To An End"[324]. *Aviation Week & Space Technology*. An epic journey of discovery at Saturn ends, leaving mysteries for future explorers.

External links

> Wikimedia Commons has media related to *Cassini-Huygens*.

Official websites

- *Cassini-Huygens* website[307] by the Jet Propulsion Laboratory
- *Cassini-Huygens* website[325] by NASA
- *Cassini-Huygens* website[308] by the European Space Agency
- *Cassini-Huygens* website[326] by NASA's Solar System Exploration division

Media and telecommunications

- CICLOPS.org[327], *Cassini* imaging homepage
- *Cassini* Hall of Fame[328], image galleries by the Jet Propulsion Laboratory
- "Cassini at Saturn"[329], a YouTube playlist by the Jet Propulsion Laboratory
  - "Titan Touchdown"[330], Depiction of *Huygens* descent and landing
- DESCANSO DSN Telecom information[331]
- *In Saturn's Rings*[332], film animated from millions of still photographs
- *Around Saturn*[333], film animated from more than 200,000 images taken by *Cassini* from 2004 to 2012
- WebGL-based 3D rendering of *Cassini*[334]

Appendix

References

[1]

[2] Based on the volume within the level of 1 bar atmospheric pressure

[3] Saturn http://www.universeguide.com/Saturn.aspx. Universe Guide. Retrieved 29 March 2009.

[4] Bizarre geometric shapes that appear at the center of swirling vortices in planetary atmospheres might be explained by a simple experiment with a bucket of water but correlating this to Saturn's pattern is by no means certain.

[5] Laboratory experiment of spinning disks in a liquid solution forms vortices around a stable hexagonal pattern similar to that of Saturn's.

[6] Jean Meeus, *Astronomical Algorithms* (Richmond, VA: Willmann-Bell, 1998). Average of the nine extremes on p 273. All are within 0.02 AU of the averages.

[7] See also the Greek article about the planet.

[8] //www.worldcat.org/issn/0027-9358

[9] //www.worldcat.org/oclc/643483454

[10] http://adsabs.harvard.edu/abs/2007CeMDA..98..155S

[11] //doi.org/10.1007/s10569-007-9072-y

[12] https://books.google.com/books?id=stFpBgAAQBAJ&pg=PA250

[13] https://solarsystem.nasa.gov/planets/saturn/overview/

[14] https://nssdc.gsfc.nasa.gov/planetary/factsheet/saturnfact.html

[15] https://planetarynames.wr.usgs.gov/Page/SATURN/system

[16] https://saturn.jpl.nasa.gov/

[17] http://solarviews.com/eng/saturn.htm

[18] https://www.youtube.com/watch?v=zVScPy-fiGM

[19] http://www.ciclops.org/view_event/185/SATURN_REV_175_RAW_PREVIEW

[20] http://www.universetoday.com/98699/saturns-strange-hexagon-in-living-color/

[21] http://photojournal.jpl.nasa.gov/jpeg/PIA10486.jpg

[22] http://photojournal.jpl.nasa.gov/targetFamily/Saturn

[23] http://apod.nasa.gov/apod/ap120122.html

[24] http://apod.nasa.gov/apod/ap121204.html

[25] http://photojournal.jpl.nasa.gov/archive/PIA11682.mov

[26] http://www.jpl.nasa.gov/news/news.php?release=2013-350

[27] http://www.planetary.org/multimedia/space-images/saturn/saturns-north-polar-vortex-animation.html

[28] http://www.ciclops.org////view_media.php?id=38331

[29] https://www.youtube.com/watch?v=VQzLY17ncWM

[30] http://www.jpl.nasa.gov/spaceimages/details.php?id=PIA21049

[31] Smith, 1959

[32] Brown, 1975

[33] Kivelson, 2005, p. 2077

[34]

[35]

[36]

[37]

[38] The subsolar point is a point on a planet, never fixed, at which the Sun appears directly overhead.

[39]

[40] Gombosi, 2009, p. 247

[41]

[42]

[43]

[44] Andre, 2008, pp. 10–15

45

[46] Mauk, 2009, pp. 317–318
[47] Gombosi, 2009, pp. 231–234
48

[49] Blanc, 2005, pp. 264–273
50

[51] On the dayside a noticeable magnetodisk only forms when the Solar wind pressure is low, and the magnetosphere has a size larger than about 23 R_s. However, when the magnetosphere is compressed by the Solar wind the dayside magnetodisk is quite small. On the other hand, in the dawn sector of the magnetosphere the disk-like configuration is present permanently.
[52] Mauk, 2009, pp. 282–283
53

[54] Young, 2005
[55] Smith, 2008
56

57

[58] Smith, 2008, pp. 1–2
[59] Gombosi, 2009, pp. 219–220
[60] Gombosi, 2009, pp. 206, 215–216
61

62

[63] Bunce, 2008, pp. 1–2
[64] The contribution of the plasma thermal pressure gradient force may also be significant.<ref name=Gombosi225>Gombosi, 2009, pp. 225–231
[65] Bunce, 2008, p. 20
[66] Kurth, 2009, pp. 334–342
67

[68] Nichols, 2009
[69] The difference between the southern and northern aurorae is related to the shift of the internal magnetic dipole to the northern hemisphere—the magnetic field in the northern hemisphere is slightly stronger than in the southern one.<ref name=Gombosi209-211>Gombosi, 2009, pp. 209–211
[70] Kurth, 2009, pp. 335–336
[71] Cowley, 2008, pp. 2627–2628
72

73

74

[75] Zarka, 2007
[76] Gurnett, 2005, p. 1256
77

78

[79] Zarka, 2005, pp. 384–385
80

[81] Mauk, 2009, pp. 286–289
[82] Leisner, 2007
83

[84] Mauk, 2009, pp. 283–284, 286–287
[85] Mauk, 2009, pp. 293–296
[86] Mauk, 2009, pp. 285–286
[87] Johnson, 2008, pp. 393–394
[88] Zarka, 2005, p. 372
[89] http://adsabs.harvard.edu/abs/2008RvGeo..46.4008A
[90] //doi.org/10.1029/2007RG000238
[91] http://hal.archives-ouvertes.fr/docs/00/31/80/22/PDF/angeo-24-1145-2006.pdf
[92] http://adsabs.harvard.edu/abs/2006AnGeo..24.1145B
[93] //doi.org/10.5194/angeo-24-1145-2006

[94] http://www.bu.edu/csp/uv/cp-aeronomy/Bhardwaj_Gladstone_RG_2000.pdf
[95] http://adsabs.harvard.edu/abs/2000RvGeo..38..295B
[96] //doi.org/10.1029/1998RG000046
[97] http://adsabs.harvard.edu/abs/2005SSRv..116..227B
[98] //doi.org/10.1007/s11214-005-1958-y
[99] http://adsabs.harvard.edu/abs/1975ApJ...198L..89B
[100] //doi.org/10.1086/181819
[101] http://www-ssc.igpp.ucla.edu/personnel/russell/papers/CassiniObservations.pdf
[102] http://adsabs.harvard.edu/abs/2007JGRA..11210202B
[103] //doi.org/10.1029/2007JA012275
[104] https://web.archive.org/web/20110716214344/http://www.thaispaceweather.com/BU2005.pdf
[105] http://adsabs.harvard.edu/abs/2005Natur.433..717C
[106] //doi.org/10.1038/nature03331
[107] //www.ncbi.nlm.nih.gov/pubmed/15716945
[108] http://www.thaispaceweather.com/BU2005.pdf
[109] http://www.ann-geophys.net/26/2613/2008/angeo-26-2613-2008.html
[110] http://adsabs.harvard.edu/abs/2008AnGeo..26.2613C
[111] //doi.org/10.5194/angeo-26-2613-2008
[112] //doi.org/10.1007/978-1-4020-9217-6_9
[113] http://adsabs.harvard.edu/abs/2005Sci...307.1255G
[114] //doi.org/10.1126/science.1105356
[115] //www.ncbi.nlm.nih.gov/pubmed/15604362
[116] http://people.virginia.edu/~rej/papers06/johnson-icarus06.pdf
[117] http://adsabs.harvard.edu/abs/2006Icar..180..393J
[118] //doi.org/10.1016/j.icarus.2005.08.021
[119] http://www.igpp.ucla.edu/people/mkivelson/Publications/285-SSR11629905.pdf
[120] http://adsabs.harvard.edu/abs/2005SSRv..116..299K
[121] //doi.org/10.1007/s11214-005-1959-x
[122] http://www.igpp.ucla.edu/people/mkivelson/Publications/287-ASR362077.pdf
[123] http://adsabs.harvard.edu/abs/2005AdSpR..36.2077K
[124] //doi.org/10.1016/j.asr.2005.05.104
[125] //doi.org/10.1007/978-1-4020-9217-6_12
[126] http://adsabs.harvard.edu/abs/2007LPI....38.1425L
[127] //doi.org/10.1007/978-1-4020-9217-6_11
[128] http://hubblesite.org/pubinfo/pdf/2010/09/pdf.pdf
[129] http://adsabs.harvard.edu/abs/2009GeoRL..3624102N
[130] //doi.org/10.1029/2009GL041491
[131] http://people.virginia.edu/~rej/papers08/paranicas-icarus08.pdf
[132] http://adsabs.harvard.edu/abs/2008Icar..197..519P
[133] //doi.org/10.1016/j.icarus.2008.05.011
[134] http://www.iop.org/EJ/article/0034-4885/56/6/001/rp930601.pdf
[135] http://adsabs.harvard.edu/abs/1993RPPh...56..687R
[136] //doi.org/10.1088/0034-4885/56/6/001
[137] http://www-ssc.igpp.ucla.edu/personnel/russell/papers/Titan%27s_influence_Saturnian_substorm_occurrence.pdf
[138] http://adsabs.harvard.edu/abs/2008GeoRL..3512105R
[139] //doi.org/10.1029/2008GL034080
[140] http://people.virginia.edu/~rej/papers08/sittlerPSS07.pdf
[141] http://adsabs.harvard.edu/abs/2008P&SS...56....3S
[142] //doi.org/10.1016/j.pss.2007.06.006
[143] http://people.virginia.edu/~rej/papers09/smithetal-JGR09.pdf
[144] http://adsabs.harvard.edu/abs/2008JGRA..11311206S
[145] //doi.org/10.1029/2008JA013352
[146] http://adsabs.harvard.edu/abs/1959ApJ...130..641S
[147] //doi.org/10.1086/146753
[148] http://adsabs.harvard.edu/abs/2006Sci...311.1409T

[149] //doi.org/10.1126/science.1121061

[150] //www.ncbi.nlm.nih.gov/pubmed/16527967

[151] http://adsabs.harvard.edu/abs/2005Sci...307.1262Y

[152] //doi.org/10.1126/science.1106151

[153] //www.ncbi.nlm.nih.gov/pubmed/15731443

[154] http://adsabs.harvard.edu/abs/2005SSRv..116..371Z

[155] //doi.org/10.1007/s11214-005-1962-2

[156] https://web.archive.org/web/20110603235629/http://www.obspm.fr/actual/nouvelle/nov07/zarka_etal_nature_2007.pdf

[157] http://adsabs.harvard.edu/abs/2007Natur.450..265Z

[158] //doi.org/10.1038/nature06237

[159] //www.ncbi.nlm.nih.gov/pubmed/17994092

[160] http://www.obspm.fr/actual/nouvelle/nov07/zarka_etal_nature_2007.pdf

[161] http://www-ssc.igpp.ucla.edu/personnel/russell/papers/mass_saturn%27s_magnetodisc.pdf

[162] http://adsabs.harvard.edu/abs/2007GeoRL..3409108A

[163] //doi.org/10.1029/2006GL028921

[164] http://caps.space.swri.edu/caps/teamMeetings/MtgMinutes32/TM32_MBurger_water_escape.pdf

[165] http://adsabs.harvard.edu/abs/2007JGRA..112.6219B

[166] //doi.org/10.1029/2006JA012086

[167] http://www-ssc.igpp.ucla.edu/personnel/russell/papers/plasmoids_saturn.pdf

[168] http://adsabs.harvard.edu/abs/2008JGRA..11301214H

[169] //doi.org/10.1029/2007JA012626

[170] http://space.ustc.edu.cn/cforums/notice/20070502185313.879/20071221141850.360/at/a%20dynamic,%20rotating%20ring%20current%20around%20saturn.pdf

[171] http://adsabs.harvard.edu/abs/2007Natur.450.1050K

[172] //doi.org/10.1038/nature06425

[173] //www.ncbi.nlm.nih.gov/pubmed/18075586

[174] http://people.virginia.edu/~rej/papers08/2008GL035433.pdf

[175] http://adsabs.harvard.edu/abs/2008GeoRL..3520103M

[176] //doi.org/10.1029/2008GL035433

[177] http://www-ssc.igpp.ucla.edu/personnel/russell/papers/magnetospheres_jupiter_saturn.pdf

[178] http://adsabs.harvard.edu/abs/2008AdSpR..41.1310R

[179] //doi.org/10.1016/j.asr.2007.07.037

[180] http://people.virginia.edu/~rej/papers07/smith07.pdf

[181] http://adsabs.harvard.edu/abs/2007Icar..188..356S

[182] //doi.org/10.1016/j.icarus.2006.12.007

[183] http://www.igpp.ucla.edu/people/mkivelson/Publications/305-2007JA012254.pdf

[184] http://adsabs.harvard.edu/abs/2007JGRA..11212222S

[185] //doi.org/10.1029/2007JA012254

[186] http://adsabs.harvard.edu/abs/2008Natur.453.1083S

[187] //doi.org/10.1038/nature07077

[188] //www.ncbi.nlm.nih.gov/pubmed/18563160

[189] http://saturn.jpl.nasa.gov/news/cassinifeatures/feature20110322/

[190] http://www.nasa.gov/mission_pages/cassini/multimedia/pia07966.html

[191] The mass of the rings is about the mass of Mimas, whereas the combined mass of Janus, Hyperion and Phoebe—the most massive of the remaining moons—is about one-third of that. The total mass of the rings and small moons is around 5.5×10^{19} kg.

[192] Lakdawalla, E. 2012.

[193] Inktomi was once known as "The Splat".

[194] The photometric color may be used as a proxy for the chemical composition of satellites' surfaces.

[195] A confirmed moon is given a permanent designation by the IAU consisting of a name and a Roman numeral. The nine moons that were known before 1900 (of which Phoebe is the only irregular) are numbered in order of their distance from Saturn; the rest are numbered in the order by which

they received their permanent designations. Nine small moons of the Norse group and S/2009 S 1 have not yet received a permanent designation.

[196] The diameters and dimensions of the inner moons from Pan through Janus, Methone, Pallene, Telepso, Calypso, Helene, Hyperion and Phoebe were taken from Thomas 2010, Table 3. Diameters and dimensions of Mimas, Enceladus, Tethys, Dione, Rhea and Iapetus are from Thomas 2010, Table 1. The approximate sizes of other satellites are from the website of Scott Sheppard.

[197] Masses of the large moons were taken from Jacobson, 2006. Masses of Pan, Daphnis, Atlas, Prometheus, Pandora, Epimetheus, Janus, Hyperion and Phoebe were taken from Thomas, 2010, Table 3. Masses of other small moons were calculated assuming a density of 1.3 g/cm^3.

[198] The orbital parameters were taken from Spitale, et al. 2006, IAU-MPC Natural Satellites Ephemeris Service, and NASA/NSSDC.

[199] Negative orbital periods indicate a retrograde orbit around Saturn (opposite to the planet's rotation).

[200] To Saturn's equator for the regular satellites, and to the ecliptic for the irregular satellites

[201] S/2004 S 4 was most likely a transient clump—it has not been recovered since the first sighting.

[202] E. Asphaug and A. Reufer. Middle sized moons as a consequence of Titan's accretion. Icarus.

[203] http://home.dtm.ciw.edu/users/sheppard/satellites/satsatdata.html

[204] https://web.archive.org/web/20110823202308/http://orinetz.com/planet/animatesystem.php?ephid=Q07IAL5QATR7V073RO44XQVPA00001

[205] http://orinetz.com/planet/animatesystem.php?ephid=Q07IAL5QATR7V073RO44XQVPA00001

[206] https://web.archive.org/web/20100527141504/http://solarsystem.nasa.gov/planets/profile.cfm?Object=Saturn&Display=Rings

[207] http://solarsystem.nasa.gov/planets/profile.cfm?Object=Saturn&Display=Rings

[208] http://www.astronomycast.com/astronomy/episode-61-saturns-moons/

[209] https://www.youtube.com/watch?v=xxXa9pxwzoY

[210] http://www.gethow.org/the-top-10-largest-planetary-moons

[211] https://www.nytimes.com/interactive/2015/12/18/science/space/nasa-cassini-maps-saturns-moons.html

[212] http://www.planetary.org/blogs/emily-lakdawalla/2017/0517-saturns-small-satellites-to-scale.html

[213] Jerome Brainerd, "Saturn's Rings" http://www.astrophysicsspectator.com/topics/planets/SaturnRings.html, *The Astrophysics Spectator*, Issue 1.8, 24 November 2004, retrieved 27 May 2009.

[214] http://planetarynames.wr.usgs.gov/append8.html

[215] http://nssdc.gsfc.nasa.gov/planetary/factsheet/satringfact.html

[216] Harland, David M., *Mission to Saturn: Cassini and the Huygens Probe*, Chichester: Praxis Publishing, 2002.

[217] Archie Frederick Collins, The greatest eye in the world: astronomical telescopes and their stories, page 8

[218] http://www.cbat.eps.harvard.edu/iauc/08400/08401.html

[219] http://www.cbat.eps.harvard.edu/iauc/08400/08432.html

[220] http://ciclops.org/view.php?id=3806

[221] NASA Planetary Photojournal PIA08328: Moon-Made Rings http://photojournal.jpl.nasa.gov/catalog/PIA08328

[222] Schenk Hamilton et al. 2011, pp. 751–53.

[223] Mason 2010.

[224] NASA Space Telescope Discovers Largest Ring Around Saturn http://www.jpl.nasa.gov/news/news.cfm?release=2009-150

[225] http://photojournal.jpl.nasa.gov/catalog/PIA09883

[226] https://web.archive.org/web/20050826083057/http://pds-rings.seti.org/saturn/saturn.html

[227] https://web.archive.org/web/20100527141504/http://solarsystem.nasa.gov/planets/profile.cfm?Object=Saturn&Display=Rings

[228] http://solarsystem.nasa.gov

[229] http://planetarynames.wr.usgs.gov/Page/Rings#saturn

[230] http://planetarynames.wr.usgs.gov

231 https://www.space.com/29624-giant-saturn-ring-even-bigger.html
232 http://www.bridgingthegaps.ie/2017/04/planetary_ring_systems_with_dr_mark_showalter/
233 https://www.flickr.com/photos/136797589@N04/34335296582/
234 B. Liou Gilles "Naissance de la ligue latine. Mythe et cult de fondation" in *Revue belge de philologie et d'histoire* **74** 1. pp. 73-97: 75. Citing Festus s.v. Saturnia p. 430 L: "Saturnia Italia, et mons, qui nunc est Capitolinus, Saturnius appellabatur, quod in tutela Saturni esse existimantur. Saturnii quoque dicebantur, qui castrum in imo clivo Capitolino incolebant, ubi ara diacata ei deo ante bellum troianum videtur." And Ovid "Fasti" 6, 31: "A patre dicta meo quondam Saturnia Roma est".
235 *Saturni filius*, frg. 2 in the edition of Baehrens.
236 Hans Friedrich Mueller, "Saturn," *Oxford Encyclopedia of Ancient Greece and Rome* (Oxford University Press, 2010), p. 222.
237 G. Dumézil p. 244.
238 G. Dumézil "Lua Mater" in *Déesses latines et mythes védiques* Bruxelles 1968 1959 p. 98-115. D. compared this Roman figure with Indian deity Nírṛti.
239 Varro, *De lingua latina* 5.64.
240 CIL I 2nd 449.
241 Briquel p. 144.
242 G. Alessio "Genti e favelle dell' antica Apulia" Taranto Cressati 1949 = *Archivio Storico Pugliese* 1949 II 1 p. 14.
243 Macrobius, *Saturnalia* 1.7.25.
244 Frederick Kaufman in *Harper's Magazine*, February 2008: *Wasteland. A journey through the American cloaca.*
245 Festus, *De verborum significatu* 432L.
246 Varro *Lingua Latina* V 52.
247 Mueller, "Saturn," p. 222.
248 In *Mélanges A. Grenier* Bruxelles 1962 p. 757-762 as cited by Briquel p. 141.
249 Samuel L. Macy, entry on "Father Time," in *Encyclopedia of Time* (Taylor & Francis, 1994), pp. 208–209.
250 G. Dumézil *La religion romain archaïque* Paris 1974 part I chap. 5; Italian translation Milan 1977 p. 244-245.
251 D. Briquel "Jupiter, Saturn et le Capitol. Essai de comparaison indoeuropéenne" in *Revue de l'histoire des religions* **198** 2 1981 p. 131-162. A. Brelich *Tre variazioni romane sul tema delle origini* Roma 1956. G. Piccaluga *Terminus* Roma 1974.
252 Briquel 1981 pp. 142 ff.
253 Macrobius *Saturnalia* I 7. Cited by Briquel above p. 143.
254 Iuventas shows a clear Varunian character in the liaison of Romulus with the *iuvenes* the young soldiers; Terminus has a Mitran character even though he shows Varunian traits in allowing the enlargement of the borders (*propagatio finium*): Briquel p. 134 n. 8.
255 G. Dumézil *Mitra-Varuna* Paris 1940; *Les dieux souverains des Indo-Européens* Paris 1977.
256 Briquel p. 151 citing Pliny II 138-139; Servius *Ad Aeneidem* I 42; XI 259. Saturn's lightning-bolts are those of wintertime.
257 Dionysius of Halicarnassus *Roman Anitiquities* I.19.1; Macrobius *Saturnalia* I.7.27–31
258 Ovid *Fasti* V.621–662, particularly 626–629.
259 Briquel p. 148 who cites Servius *Ad Aenaeidem* III 407.
260 Briquel p. 148 n. 63 who cites Plutarch *Quaestiones Romanae* II.
261 Pliny, *Natural History* 15.32.
262 Macrobius, *Saturnalia* 1.8.5.
263 Robert Graves, "The Greek Myths: 1" page 41
264 Tertullian, *De testimonio animae* 2.
265 Dionysius of Halicarnassus, *Antiquitates Romanae* 7.72.13.
266 Briquel p. 155.
267 Found on the Piacenza Liver; see also Martianus Capella 1.58. Mueller, "Saturn," p. 222.
268 Mueller, "Saturn," p. 221.
269 M. Leglay *Saturn africaine. Histoire* Paris BEFRA 1966.
270 Leglay p. 385-386.

[271] William F. Hansen, *Ariadne's Thread: A Guide to International Tales Found in Classical Literature* (Cornell University Press, 2002), p. 385.

[272] Macrobius, *Saturnalia* 1.1.8–9; Jane Chance, *Medieval Mythography: From Roman North Africa to the School of Chartres, A.D. 433–1177* (University Press of Florida, 1994), p. 71.

[273] see Robert A. Kaster, *Macrobius: Saturnalia, Books 1–2* (Loeb Classical Library, 2011), note on p. 16.

[274] Macrobius *Saturnalia* I 7, 18.

[275] Macrobius Saturnalia I, 9; Vergil Aeneis VII, 49

[276] Varro *Lingua Latina* V 42 and 45; Vergil *Aeneis* VIII 357-8; Dionysius of Halicarnassus *Roman Antiquities* I 34; Festus p. 322 L; Macrobius *Sat.* I 7, 27 and I 10, 4; Pliny the Elder *Natural History* III 68; Minucius Felix *Octavius* 22; Tertullian *Apologeticum* 10 as cited by Briquel p. 154.

[277] Versnel, "Saturnus and the Saturnalia," pp. 138–139.

[278] Versnel, "Saturnus and the Saturnalia," p. 139. The Roman theologian Varro listed Saturn among the Sabine gods.

[279] Versnel, "Saturnus and the Saturnalia," pp. 139, 142–143.

[280] Versnel, "Saturnus and the Saturnalia," p. 143.

[281] Virgil, *Aeneid* 8.320–325, as cited by Versnel, "Saturnus and the Saturnalia," p. 143.

[282] Mario Pincherle, Giuliana C. Volpi: *La civiltà minoica in Italia. Le città saturnie* Pisa Pacini 1990.

[283] Mueller, "Saturn," in *The Oxford Encyclopedia of Ancient Greece and Rome*, p. 222.

[284] Versnel, "Saturnus and the Saturnalia," p. 144.

[285] H.S. Versnel, "Saturnus and the Saturnalia," in *Inconsistencies in Greek and Roman Religion: Transition and Reversal in Myth and Ritual* (Brill, 1993, 1994), pp. 144–145. See also the Etruscan god Satre.

[286] For instance, Ausonius, *Eclogue* 23 and *De feriis Romanis* 33–7. See Versnel, "Saturnus and the Saturnalia," pp. 146 and 211–212, and Thomas E.J. Wiedemann, *Emperors and Gladiators* (Routledge, 1992, 1995), p. 47.

[287] Eight days were subsidized from the Imperial treasury *(arca fisci)*, and two mostly by the sponsoring magistrate himself; Michele Renee Salzman, *On Roman Time: The Codex-Calendar of 354 and the Rhythms of Urban Life in Late Antiquity* (University of California Press, 1990), p. 186.

[288] Mueller, "Saturn," in *The Oxford Encyclopedia of Ancient Greece and Rome*, p. 222; Versnel, "Saturnus and the Saturnalia," p. 146.

[289] Macrobius, *Saturnalia* 1.7.31; Versnel, "Saturnus and the Saturnalia," p. 146.

[290] Macrobius, *Saturnalia* 1.10.24; Carlin A. Barton, *The Sorrows of the Ancient Romans: The Gladiator and the Monster* (Princeton University Press, 1993), p. 166. For other Roman practices that may represent substitutes for human sacrifice, see Argei and oscilla, the latter of which were used also at the Latin Festival and the Compitalia: William Warde Fowler, *The Roman Festivals of the Period of the Republic* (London, 1908), p. 272.

[291] Versnel, "Saturnus and the Saturnalia," p. 162.

[292] http://warburg.sas.ac.uk/vpc/VPC_search/subcats.php?cat_1=5&cat_2=182

[293] http://www.wdl.org/en/item/2997/

[294] http://spaceprojects.arc.nasa.gov/Space_Projects/pioneer/PN10&11.html

[295] Cassini–Huygens: News http://saturn.jpl.nasa.gov/news/press-release-details.cfm?newsID=639

[296] New Ring Spotted Around Saturn http://www.cnn.com/2006/TECH/space/09/20/saturn.ring.reut/index.html – Article on CNN.com.

[297] *Cassini* Solstice Mission http://saturn.jpl.nasa.gov/mission/introduction/

[298] http://saturn.jpl.nasa.gov

[299] http://sci.esa.int/science-e/www/area/index.cfm?fareaid=106

[300] NASA and ESA Prioritize Outer Planet Missions http://www.nasa.gov/topics/solarsystem/features/20090218.html

[301] Jupiter in space agencies' sights http://news.bbc.co.uk/1/hi/sci/tech/7897585.stm

[302] Enceladus Life Finder http://www.hou.usra.edu/meetings/lpsc2015/pdf/1525.pdf 2015, PDF.

[303] https://books.google.com/books?id=rn7vAAAAMAAJ

[304] http://voyager.jpl.nasa.gov/science/saturn.html
[305] https://web.archive.org/web/20070426162116/http://saturn.jpl.nasa.gov/home/index.cfm
[306] http://nssdc.gsfc.nasa.gov/nmc/spacecraftDisplay.do?id=1997-061A
[307] https://saturn.jpl.nasa.gov/
[308] http://www.esa.int/Our_Activities/Space_Science/Cassini-Huygens
[309] http://www.asi.it/en/activity/solar-system-exploration/cassini-huygens
[310] https://www.jpl.nasa.gov/video/details.php?id=1468
[311] Michael W Leeds: AIAA 96-2864 Development of the Cassini Propulsion Subsystem https://trs.jpl.nasa.gov/bitstream/handle/2014/26073/96-1093.pdf?sequence=1&isAllowed=y. 32nd AIAA/ASME/SAE/ASEE Joint Propulsion Conference, July 1, 1996, retrieved January 8, 2016
[312] Ruslan Krivobok: Russia to develop nuclear-powered spacecraft for Mars mission http://en.rian.ru/analysis/20091111/156797969.html. Ria Novosti, November 11, 2009, retrieved January 2, 2011
[313] Flux and composition of interstellar dust at Saturn from Cassini's Cosmic Dust Analyzer http://science.sciencemag.org/content/352/6283/312. N. Altobelli1, F. Postberg, K. Fiege, M. Trieloff, H. Kimura, V. J. Sterken. *Science* April 15, 2016: Vol. 352, Issue 6283, pp. 312-318.
[314] *Cassini* Final Environmental Impact Statement http://saturn.jpl.nasa.gov/spacecraft/safety/safetyeis/ , see Chapter 2 http://saturn.jpl.nasa.gov/spacecraft/safety/chap2.pdf , Table 2-8
[315] How to Land on Titan http://www.ingenia.org.uk/ingenia/articles.aspx?Index=317, *Ingenia*, June 2005
[316] How Huygens avoided disaster http://www.thespacereview.com/article/306/1, James Oberg, *The Space Review*, January 17, 2005.
[317] This is currently the best measurement of post-Newtonian parameter γ; the result $\gamma = 1 + (2.1 \pm 2.3) \times 10^{-5}$ agrees with the prediction of standard General Relativity, $\gamma = 1$
[318] Meltzer 2015, pp. 346-351
[319] http://esamultimedia.esa.int/docs/titanraw/index.htm
[320] *Cassini* Spacecraft to Dive Into Water Plume of Saturn Moon http://www.nasa.gov/mission_pages/cassini/media/cassini-20080310.html NASA.gov, March 10, 2008
[321] *Cassini* Tastes Organic Material at Saturn's Geyser Moon http://www.nasa.gov/mission_pages/cassini/media/cassini-20080326.html NASA, March 26, 2008
[322] NASA Extends Cassini's Tour of Saturn, Continuing International Cooperation for World Class Science http://saturn.jpl.nasa.gov/news/newsreleases/newsrelease20100203/. NASA / California Institute of Technology / Jet Propulsion Laboratory, February 3, 2010, retrieved January 2, 2011
[323] http://ciclops.org/index.php
[324] http://aviationweek.com/space/cassini-s-ringside-seat-saturn-coming-end
[325] https://www.nasa.gov/cassini
[326] https://solarsystem.nasa.gov/missions/cassini
[327] http://ciclops.org/
[328] https://saturn.jpl.nasa.gov/galleries/hall-of-fame/
[329] https://www.youtube.com/playlist?list=PLTiv_XWHnOZpKPaDTVy36z0U8GxoiIkZa
[330] https://www.youtube.com/watch?v=msiLWxDayuA
[331] https://descanso.jpl.nasa.gov/DPSummary/Descanso3--Cassini2.pdf
[332] http://insaturnsrings.com/
[333] https://vimeo.com/70532693
[334] http://spacecrafts3d.org/models/cassini.html

Article Sources and Contributors

The sources listed for each article provide more detailed licensing information including the copyright status, the copyright owner, and the license conditions.

Saturn *Source:* https://en.wikipedia.org/w/index.php?oldid=856189818 *License:* Creative Commons Attribution-Share Alike 3.0 *Contributors:* 564dude, 72, A Great Catholic Person, A2soup, Adam9007, Ahpook, AlexiusHoratius, Archeologo, Arjayay, Awesomegamer, BD2412, BatteryIncluded, Bender235, Bentogoa, Bobogoobo, Brandmeister, C.Fred, Ceoil, Churri101, ClueBot NG, ComicsAreJustAllRight, Crystallizedcarbon, Cyrus noto3at bulaga, D.C.Rigate, DavideVeloria88, Dawnseeker2000, Dewritech, Dogman15, Donner60, Double sharp, DrKay, Drbogdan, EP111, ESkog, Edulovers, Edwininlondon, Ehrenkater, El C, Eurodyne, Ewanmellor, Excirial, Fbergo, FlightTime, Flyer22 Reborn, Frood, GSS, GreenMeansGo, Gulumeemee, Gymnophoria, HalloweenNight, Hamiltondaniel, Hansonjay, Hdjensofjfnen, Headbomb, Hellrider19999, Hillbillyholiday, Hunterm267, Huntster, Iggy the Swan, In ictu oculi, Isambard Kingdom, JRPG, JamesCox 13, Jimfbleak, JoeHebda, John, John of Reading, Jon Kolbert, JorisvS, JustinTime55, KH-1, Kalimantanas, Kanjuzi, Keith D, Kozan Huseyin, L3X1, LC03, Labtek00, Larry Hockett, Lasunncty, LiberatorG, Loodog, Lornof, Mahveotm, MartinZ, Materialscientist, Mfb, Miles Edgeworth, Mojoworker, Niceguyedc, Nikkimaria, Noscoper21, Octoberwoodland, Omanyd, Oshwah, Paintspot, Parejkoj, Pdebee, PhilipTerryGraham, Piledhigheranddeeper, PlanetUser, Pluma, Praemonitus, Randy Kryn, Redav, Rfassbind, Rjwilmsi, Ruslik0, Sakura Cartelet, Samsara, Saros136, Sebastian.318, Seleepen, Serols, Sharfooth, Sheldon4d, Silverhorse, Simplexity22, SkyWarrior in public, Space-Age Meat, SparklingPessimist, Spartanwolf223, SpeedEvil, Steve03Mills, StewartIM, SubSpace, TAnthony, Tbbotch, The Transhumanist, Tom.Reding, TomGeog, Tompop888, TonyBallioni, Twilight Magic, TwoTwoHello, Unbuttered Parsnip, User-duck, VT0207, VVsuace666, Voello, Vsmith, W like wiki, Wedemboyz2088, Whack boy 2.0, WolfmanSF, Worldbruce, Writerwhiz2014, Zavierpl, Zedshort, 111 anonymous edits 1

Saturn's hexagon *Source:* https://en.wikipedia.org/w/index.php?oldid=852992481 *License:* Creative Commons Attribution-Share Alike 3.0 *Contributors:* Alexf, Andy Dingley, Artman40, Bearcat, Benjammin105123, Bgwhite, BkDJk, BoxOfChickens, CES1596, Canterbury Tail, Chevy111, Chris Capoccia, ClueBot NG, Dawnseeker2000, Deacon Vorbis, DivineAlpha, Drbogdan, Drpickem, Eanpowell, Fotaun, Goustien, Hike395, Howpper, JMtB03, JamesEG, Jcrocker, Jeffq, Jerzy, John, John11235813, Jon Jonathan, JorisvS, Jpgordon, Kyleleitch, LittleWink, Lumos3, Magioladitis, Malcolmredheron, Maplestrip, Marcus Cyron, Marek69, Martarius, Materialscientist, Megantcflux, MereTechnicality, Merpings, No Swan So Fine, Oshwah, PhilipTerryGraham, Pifactorial, RagingR2, Redrose64, Rjwilmsi, Sae1962, Sidelight12, Silenceisgod, Stareditor88, Stewart Little, Stikkyy, Stub Mandrel, Thomas.W, Tillman, Tom mai78101, Tom.Reding, Travelbird, Trevayne08, Wilkyway7, WolfmanSF, XavierGreen, Xx asia 10101lllll, Ziggaway, Zveznet, 54 anonymous edits 25

Magnetosphere of Saturn *Source:* https://en.wikipedia.org/w/index.php?oldid=855683569 *License:* Creative Commons Attribution-Share Alike 3.0 *Contributors:* Aarghdvaark, Ajsteffl, Anthonyhcole, ArchonMagnus, Citation bot 1, ClueBot NG, Corginaut, Dawnseeker2000, Deneb in Cygnus, DocWatson42, Dodshe, Double sharp, Drbogdan, Drilnoth, El C, Elfguy, Eyreland, Gaius Cornelius, Glane23, Headbomb, Huntster, Hydraton31, Ixfd64, JaconaFrere, JorisvS, Keithh, Kolbasz, Kwamikagami, LilHelpa, Mild Bill Hiccup, Newone, Originalwana, Picklehead101, Pjoef, Remember, Riddleh, Rjwilmsi, Ruslik0, Sardanaphalus, SentientSystem, Serendipodous, Serols, Skizzik, Steel1943, Tabletop, Tassedethe, Tesscass, The High Fin Sperm Whale, Tom.Reding, Viriditas, WolfmanSF, Woohookitty, 36 anonymous edits 29

Moons of Saturn *Source:* https://en.wikipedia.org/w/index.php?oldid=855292909 *License:* Creative Commons Attribution-Share Alike 3.0 *Contributors:* 180-over-Pi, 5 albert square, 65HCA7, 7Sidz, A2soup, Ahonc, AntanO, ArnoldReinhold, Astrokat, Babycakes4444444000, Beauty School Dropout, Begoon, Bender235, Bmclaughlin9, CambridgeBayWeather, CielProfond, ClueBot NG, CommonsDelinker, CubeSat4U, CuriousMind01, DatGuy, David-abbott9011, Dawnseeker2000, Dcirovic, Double sharp, Drbogdan, DrunkBicyclist, Earthandmoon, F6Zman, Finnusertop, FlightTime, FlightTime Phone, Frze, GDK, Gap9551, Gfgfgfgvvv, Grayfell, H.dryad, Hacker10,000, Headbomb, I dream of horses, Ignatzmice, Ishaan9, J. 'mach' wust, JDAWiseman, Jai1971, Jim1138, Joe Decker (alt), JorisvS, Kogge, Kwamikagami, KylieTastic, LittleWink, LorenzoB, MaisNautron, Marteau, Materialscientist, Math-Keduor7, Matheus Faria, Matthewedwards, Mcmatter, Me, Myself, and I are Here, Modest Genius, Mortense, Mr KEBAB, Musa Raza, MusikAnimal, Nardog, NawlinWiki, Nergaal, Newone, Octoberwoodland, Orange Suede Sofa, Paul August, Racerx11, Reatlas, Rjwilmsi, Rothorpe, Ruslik0, Schockading, ScottM84, SemiHypercube, SenseiAC, Seththekillerx, SidneyN11, Stephenb, Stephenbyrne18, StewartIM, Stongduke, StringTheory11, Stub Mandrel, Tamfang, Tetra quark, TheLlNguy, Theinstantmatrix, Titus III, Tom.Reding, Tomruen, Trappist the monk, Tripple figgure, Wercloud, Wikifire42, Wingedsubmariner, WolfmanSF, XinaNicole, Yedwardmcinnes, Zingvin, Internion, 175 anonymous edits 47

Rings of Saturn *Source:* https://en.wikipedia.org/w/index.php?oldid=856081820 *License:* Creative Commons Attribution-Share Alike 3.0 *Contributors:* 72, Almitydave, Amend12345, AntHerder, Arado, Arjayay, BatteryIncluded, Bender235, Bosley John Bosley, Caballero1967, CanX 322, ClueBot NG, Cursed6666, DSmurf, DV0m, Dawnseeker2000, Deneb in Cygnus, Edaham, Edward Z, Exoplanetaryscience, Favonian, FlightTime, Flyer22 Reborn, Fraggle81, Frietjes, Gilliam, Grammarian79, GreatWikiBen, Hillbillyholiday, Huritisho, I dream of horses, IceKarma, ImranSomji, Infamous Castle, Ixfd64, J 1982, Jcpag2012, Jim1138, Jimgibson1, John "Hannibal" Smith, Jonesey95, JorisvS, K6ka, K9re11, Karatechopfury, Karthikrrd, Kogge, Krassotkin, LittlePuppers, MONGO, Manul, Materialscientist, Matma Rex, Maye, Mortense, Ninney, NinuKinuski, Nwbeeson, Omeganian, Ost316, PlanetUser, Randy Kryn, RichardFloyd, Rjwilmsi, Robertinventor, Rsrikanth05, Ruslik0, Seahorseruler, SemiHypercube, Serasuna, Serols, Shellwood, Simplexity22, SkyWarrior, Stamptrader, Steel1943, Stinky74965, TNTPublic, Tanishqga111, The Original Filfi, Tom.Reding, Tony North, Topbanana, Trappist the monk, Twinsday, Vermont, Volunteer1234, WOSlinker, Widr, WolfmanSF, Yowanvista, Zdp619, Zingvin, 150 anonymous edits 79

Saturn (mythology) *Source:* https://en.wikipedia.org/w/index.php?oldid=855063538 *License:* Creative Commons Attribution-Share Alike 3.0 *Contributors:* Alagos, Aldrasto11, Amcrius, Asarelah, Avoided, Ben Ammi, Bjhodge8, Black Falcon, Bloodofox, Boudicca100, Brandmeister, C.Fred, CAPTAIN RAJU, Cerabot~enwiki, Chumbough, ClueBot NG, Cmcalpine, Cynwolfe, DavideVeloria88, Dhaidziad, Dbachmann, Denisarona, Dewritech, Drbogdan, Ducknish, EamonnPKeane, Ebyabe, Edgar181, ElizaKobert, Euryalus, Favonian, Frietjes, Geekdiva, Gilliam, Gladamas, Hairhorn, Hbrown21, HouseGecko, HueSatLum, Hume42, Ibadibam, Iluvhockey, Iridescent, Jim1138, John of Reading, Jon C., Joren, Jschnur, Justlettersandnumbers, K6ka, Kathlee123, Kevin Lakhani, Kinttsubuffalo, Koornti, Lautensack, LouisAlain, Macklev, MagicatthemovieS, Manytexts, Marianna251, Michael-Zero, Missvain, Monobears, Mrchillman10, Narky Blert, Neptune's Trident, Niceguyedc, Oshwah, PackMecEng, Paul August, Peterklevy, Pinethicket, R'n'B, Rotor Army no.345, Sakura Cartelet, Serols, Sigehelmus, Sminthopsis84, Smtchahal, SpartaN, Speeditor, Sun Creator, Suslindisambiguator, Tadboy-namedalo, Tbbotch, Technopat, TheThomas, Thisisreallywrong, Timasuke, Tratchy, Vieque, WCCasey, Widr, Wutschwllim, Xiaphias, Yamaguchi先生, Zeugding, Zppix, Piudamers, 165 anonymous edits 119

Exploration of Saturn *Source:* https://en.wikipedia.org/w/index.php?oldid=850596848 *License:* Creative Commons Attribution-Share Alike 3.0 *Contributors:* 72, A7x, Amortias, Anomalocaris, Art LaPella, BatteryIncluded, BillCook, Bscuga, Charvest, ClueBot NG, Cmglee, Coreyvi66, Czolgolz, DPBT1, Dan100, Diannaa, Discospinster, Double sharp, DougsTech, Drbogdan, Elee, Entropy, Fotaun, Frosty, Gene Nygaard, Ginsuloft, Gits (Neo), IRP, JFG, Ja 62, Jonathan Hall, JorisvS, Kwamikagami, Lentower, Materialscientist, Mlm42, Petchboo, Phoenix7777, Pious7, Qwidjib0, RB10, Randy Kryn, Raymondwinn, Remember, Rich Farmbrough, Ruslik0, Sachi Bbsr, Sardanaphalus, Selolovng, Senator20229, Skizzik, Softlavender, Surajt88, Tetra quark, TheGeneralUser, TheSuave, Tom.Reding, TommyBoy, Trappist the monk, Uncle Dick, WereSpielChequers, Widr, Will Beback Auto, 83 anonymous edits 131

Cassini–Huygens *Source:* https://en.wikipedia.org/w/index.php?oldid=856457187 *License:* Creative Commons Attribution-Share Alike 3.0 *Contributors:* A D Monroe III, Alaskanloops, Aldenluke, Alethe, Alexander Davronov, Almicione, AlphaBetaGamma01, Ammarpad, AnApple159, Another Believer, Ar1001, Arjayay, Art LaPella, Atropos235, Audacity, Balon Greyjoy, BatteryIncluded, Blainster, Catostylus, Chris857, Chrisrabaya, Christ-Trekker, Ciphers, Clarities, ClueBot NG, Cmrss2, Contraption5000, Courcelles, CreationFox, Cyrus noto3at bulaga, DAVINA 6666, DRAGON BOOSTER, Damonskye, Dan100, Dashy902, Delmlsfan, Dobrichev, Dogman15, DonaldTonald3, Dondervogel 2, Double sharp, DoulosBen, Drbogdan, Drewmutt, El C, ElectricController, Entranced98, Enwebb, FT2, Fanyavizuri, Fcrary, Feldkurat Katz, Flyer22 Reborn, Fotaun, Fountains of Bryn Mawr, FrancisF23, Gemchadur, Gert Van Waelvelde, Gluonman, GrahamCracker325, Hekerui, Hmains, Huntster, HurricaneNathanMC, Ice Cap Zone, Ixfd64, J. Martin Velez Linares, JFG, Jasper Deng, Jd22292, Jge1457, Jim.henderson, Jim1138, JohnSjr, Johnster2222, Jonny2143, Jrfixer, KGirlTrucker81, KurisuYamato, L293D, Lentower, Lihaas, LilHelpa, Lucky For You, Maczkopeti, Magioladitis, Marc Lacoste, Marcocapelle, Marianna251, Masem, Materialscientist, MennasDosbin, Meno25, Mocha2007, Modest Genius, Morningstar1814, MrBruceSpringsteen, MrHarmanBoyGuy, Mssgill, NYPatel, Nergaal, NewEngland Yankee, Nixinova, Nren4237, Opencooper, Origin7, Oshwah, OwenBlacker, Parejkoj, PhilipTerryGraham, Phoenix53004, Phoenix7777, Piledhigheranddeeper, PlyrStar93, PrateekKumarDas, Randy Kryn, Ravenswing, Reality Bent, Rockhq, RoyGoldsmith, Shellwood, Simpatico qa, Simplexity22, SkyWarrior, Smashthewarontruth, Smurrayinchester, Spectaprig, StewartIM, Stratman, Suede Cat, Thayts, The Moose, TheMusicalDoc, TomCat4680, TommyBoy, Tomruen, Trong Browne, Twilight Magic, TychosElk, U-95, Ugly Ketchup, UnsungKing123, Vanamonde93, Vicsar, Volcycle, Vsmith, Wetman, What6wsd, Woodlandtrail, Wtmitchell, Xyzzyva, Yatagerasu, Zavierpl, Zedmelon, ZJames, אמא של גולן, 134 anonymous edits 137

Image Sources, Licenses and Contributors

The sources listed for each image provide more detailed licensing information including the copyright status, the copyright owner, and the license conditions.

Figure 48 *Source:* https://en.wikipedia.org/w/index.php?title=File:Backlit_Saturn_from_Cassini_Orbiter_2007_May_9.jpg *License:* Public Domain *Contributors:* 99of9, Cirt, Olybrius, PhilipTerryGraham, Pierpao, Racky, Ruslik0, Salix, Shizhao, Stas1995, The High Fin Sperm Whale, TheDJ, Tomruen, Torsch, Tryphon, WolfmanSF ... 86
Figure 49 *Source:* https://en.wikipedia.org/w/index.php?title=File:Saturn_Ring_Material.jpg *License:* Public Domain *Contributors:* NASA/JPL/University of Colorado ... 88
Image *Source:* https://en.wikipedia.org/w/index.php?title=File:Saturn's_rings_dark_side_mosaic.jpg *License:* Public Domain *Contributors:* NASA/JPL/Space Science Institute ... 90
Figure 50 *Source:* https://en.wikipedia.org/w/index.php?title=File:Saturn's_ring_plane.svg *License:* Public Domain *Contributors:* NASA/JPL/Space Science Institute ... 92
Image *Source:* https://en.wikipedia.org/w/index.php?title=File:Saturn's_rings_in_visible_light_and_radio.jpg *License:* Public Domain *Contributors:* NASA/JPL/Space Science Institute ... 93
Figure 51 *Source:* https://en.wikipedia.org/w/index.php?title=File:PIA18313_Saturn's_D_ring_and_inner_C_ring.jpg *License:* Public Domain *Contributors:* Huntster, WolfmanSF ... 93
Figure 52 *Source:* https://en.wikipedia.org/w/index.php?title=File:PIA06540_Outer_C_Ring.jpg *License:* Public Domain *Contributors:* NASA/JPL/Space Science Institute ... 94
Image *Source:* https://en.wikipedia.org/w/index.php?title=File:PIA21628_-_B_Ring_fine_structure_in_color.jpg *License:* Public Domain *Contributors:* WolfmanSF ... 96
Image *Source:* https://en.wikipedia.org/w/index.php?title=File:PIA11668_B_ring_peaks_2x_crop.jpg *License:* Public Domain *Contributors:* Huntster, WolfmanSF ... 96
Figure 53 *Source:* https://en.wikipedia.org/w/index.php?title=File:Saturn_ring_spokes_PIA11144_300px_secs15.5to23_20080926.ogv *License:* Public Domain *Contributors:* NASA/JPL/Space Science Institute ... 97
Figure 54 *Source:* https://en.wikipedia.org/w/index.php?title=File:Cassini_Division.jpg *License:* Public Domain *Contributors:* NASA/JPL/Space Science Institute ... 98
Figure 55 *Source:* https://en.wikipedia.org/w/index.php?title=File:PIA06534_Encke_Division.jpg *License:* Public Domain *Contributors:* NASA/JPL/Space Science Institute ... 99
Figure 56 *Source:* https://en.wikipedia.org/w/index.php?title=File:PIA08319_Daphnis_in_Keeler_Gap.jpg *License:* Public Domain *Contributors:* NASA/JPL/Space Science Institute ... 101
Figure 57 *Source:* https://en.wikipedia.org/w/index.php?title=File:Daphnis_edge_wave_shadows.jpg *License:* Public Domain *Contributors:* NASA/JPL/Space Science Institute ... 101
Figure 58 *Source:* https://en.wikipedia.org/w/index.php?title=File:A_Ring_propeller_"Santos-Dumont"_(PIA_21433).jpg *License:* Public Domain *Contributors:* Armbrust, PhilipTerryGraham, WolfmanSF ... 102
Figure 59 *Source:* https://en.wikipedia.org/w/index.php?title=File:First_moonlets_PIA07792.jpg *License:* Public Domain *Contributors:* NASA/JPL/Space Science Institute ... 102
Figure 60 *Source:* https://en.wikipedia.org/w/index.php?title=File:Roche_division,_rings_of_Saturn.jpg *License:* Public Domain *Contributors:* Atmoz, BotMultichill, DragonflySixtyseven, File Upload Bot (Magnus Manske), Huntster, MGA73bot2, OgreBot 2, Ruslik0, Sreejithk2000 ... 103
Figure 61 *Source:* https://en.wikipedia.org/w/index.php?title=File:PIA07712_-_F_ring_animation.ogv *License:* Public Domain *Contributors:* NASA/JPL/Space Science Institute ... 104
Image *Source:* https://en.wikipedia.org/w/index.php?title=File:F_Ring_perturbations_PIA08412.jpg *License:* Public Domain *Contributors:* Cassini Imaging Team ... 105
Figure 62 *Source:* https://en.wikipedia.org/w/index.php?title=File:Saturn_outer_rings_labeled.svg *License:* Public Domain *Contributors:* NASA/JPL/Space science institute. ... 105
Figure 63 *Source:* https://en.wikipedia.org/w/index.php?title=File:PIA11101_Anthe_ring_arc.jpg *License:* Public Domain *Contributors:* NASA / JPL / Space Science Institute ... 107
Image *Source:* https://en.wikipedia.org/w/index.php?title=File:E_ring_with_Enceladus.jpg *License:* Public Domain *Contributors:* NASA/JPL/Space Science Institute ... 108
Image *Source:* https://en.wikipedia.org/w/index.php?title=File:PIA17184_rot180_inset_PIA14658_rot38.jpg *License:* Public Domain *Contributors:* WolfmanSF ... 108
Image *Source:* https://en.wikipedia.org/w/index.php?title=File:PIA17191-SaturnMoon-Enceladus-TendrilSims-20150414.jpg *License:* Public Domain *Contributors:* Drbogdan, PhilipTerryGraham ... 109
Figure 64 *Source:* https://en.wikipedia.org/w/index.php?title=File:Infrared_Ring_Around_Saturn.jpg *License:* Public Domain *Contributors:* NASA/ESA/STScI/AURA ... 110
Figure 65 *Source:* https://en.wikipedia.org/w/index.php?title=File:PIA14943-SaturnBehindTheRings-20180813.jpg *License:* Public Domain *Contributors:* Drbogdan, Huntster, WolfmanSF ... 111
Figure 66 *Source:* https://en.wikipedia.org/w/index.php?title=File:PIA11660-_Mimas'_shadow_cut_off_by_B_ring_(trimmed).jpg *License:* Public Domain *Contributors:* NASA/JPL/Space Science Institute ... 112
Figure 67 *Source:* https://en.wikipedia.org/w/index.php?title=File:PIA21627_-_Janus_2-to-1_spiral_density_wave_in_Saturn's_inner_B_Ring.jpg *License:* Public Domain *Contributors:* OgreBot 2, WolfmanSF ... 112
Figure 68 *Source:* https://en.wikipedia.org/w/index.php?title=File:PIA22418_-_Gravity's_Rainbow_-_Saturn's_B_Ring_in_color.jpg *License:* Public Domain *Contributors:* Krassotkin, WolfmanSF ... 113
Figure 69 *Source:* https://en.wikipedia.org/w/index.php?title=File:Spokes_in_Saturn's_B_Ring.jpg *License:* Public Domain *Contributors:* NASA/JPL/Space Science Institute ... 113
Figure 70 *Source:* https://en.wikipedia.org/w/index.php?title=File:Srings.jpg *License:* Public Domain *Contributors:* NASA ... 114
Figure 71 *Source:* https://en.wikipedia.org/w/index.php?title=File:PIA06099_Enke_Gap.jpg *License:* Public Domain *Contributors:* NASA/JPL/Space Science Institute ... 114
Figure 72 *Source:* https://en.wikipedia.org/w/index.php?title=File:Daphnis_makes_waves_-_4x_vertical_stretch.jpg *License:* Public Domain *Contributors:* WolfmanSF ... 115
Figure 73 *Source:* https://en.wikipedia.org/w/index.php?title=File:Prometheus_und_Pandora.jpg *License:* Public Domain *Contributors:* NASA/JPL/Space Science Institute ... 115
Figure 74 *Source:* https://en.wikipedia.org/w/index.php?title=File:PIA12684_F_Ring.png *License:* Public Domain *Contributors:* NASA / JPL / Space Science Institute ... 116
Figure 75 *Source:* https://en.wikipedia.org/w/index.php?title=File:F_Ring_Dynamism_PIA08290.jpg *License:* Public Domain *Contributors:* Cassini Spacecraft ... 116
Figure 76 *Source:* https://en.wikipedia.org/w/index.php?title=File:PIA11635-_Slicing_the_Arc.jpg *License:* Public Domain *Contributors:* NASA/JPL/Space Science Institute ... 116
Image *Source:* https://en.wikipedia.org/w/index.php?title=File:0_Autel_dédié_au_dieu_Malakbêl_et_aux_dieux_de_Palmyra_-_Musei_Capitolini_(1b).JPG *License:* Creative Commons Attribution 3.0 *Contributors:* User:Jean-Pol GRANDMONT ... 119
Figure 77 *Source:* https://en.wikipedia.org/w/index.php?title=File:Arch_of_SeptimiusSeverus.jpg *License:* Creative Commons Attribution 2.0 *Contributors:* Robert Lowe ... 122
Figure 78 *Source:* https://en.wikipedia.org/w/index.php?title=File:Throne_of_Saturn_Louvre_Ma1662.jpg *License:* Creative Commons Attribution 2.5 *Contributors:* User:Jastrow ... 126
Figure 79 *Source:* https://en.wikipedia.org/w/index.php?title=File:Porta_Maggiore_Alatri.jpg *License:* GNU Free Documentation License *Contributors:* Attilios, BotMultichill, DenghiùComm, MGA73bot2, Torquatus ... 126
Figure 80 *Source:* https://en.wikipedia.org/w/index.php?title=File:Lucius_Appuleius_Saturninus.jpg *License:* GNU Free Documentation License *Contributors:* CNG ... 128
Image *Source:* https://en.wikipedia.org/w/index.php?title=File:Wikisource-logo.svg *License:* Creative Commons Attribution-Sharealike 3.0 *Contributors:* ChrisiPK, Guillom, INeverCry, Jarekt, JuTa, Leyo, Lokal Profil, MichaelMaggs, NielsF, Rei-artur, Rocket000, Romaine, Steinsplitter ... 129
Figure 81 *Source:* https://en.wikipedia.org/w/index.php?title=File:2009-07-21_ob_05_saturn.JPG *License:* Creative Commons Attribution-Sharealike 3.0,2.5,2.0,1.0 *Contributors:* Ziko ... 132
Figure 82 *Source:* https://en.wikipedia.org/w/index.php?title=File:P11F81.jpg *License:* Public Domain *Contributors:* BotMultichill, BotMultichillT, Conscious, Gpetrov, RadiX, Ruslik0, Shizhao, Wstrwald, 1 anonymous edits ... 133
Figure 83 *Source:* https://en.wikipedia.org/w/index.php?title=File:Saturn_eclipse.jpg *License:* Public Domain *Contributors:* NASA/JPL/Space Science Institute ... 134
Figure 84 *Source:* https://en.wikipedia.org/w/index.php?title=File:Animation_of_Cassini_trajectory_around_Saturn.gif *Contributors:* User:Phoenix7777 ... 134
Figure 85 *Source:* https://en.wikipedia.org/w/index.php?title=File:Gas_planet_size_comparisons.jpg *License:* Public Domain *Contributors:* Solar System Exploration, NASA ... 135
Image *Source:* https://en.wikipedia.org/w/index.php?title=File:Cassini_Saturn_Orbit_Insertion.jpg *License:* Public Domain *Contributors:* Glenn, Jdx, Mirecki, Moheen Reeyad, Stewi101015, Tagheuher, TheDJ, 3 anonymous edits ... 137

License

Index

www.ingramcontent.com/pod-product-compliance
Lightning Source LLC
Chambersburg PA
CBHW021925190326
41519CB00009B/917